Energy Futures

De Gruyter Contemporary Social Sciences

Volume 10

Energy Futures

Anthropocene Challenges, Emerging Technologies
and Everyday Life

Edited by
Simone Abram, Karen Waltorp, Nathalie Ortar and
Sarah Pink

DE GRUYTER

ISBN 978-3-11-162092-3
e-ISBN (PDF) 978-3-11-074564-1
e-ISBN (EPUB) 978-3-11-074569-6
ISSN 2747-5689

Library of Congress Control Number: 2022943600

Bibliographic information published by the Deutsche Nationalbibliothek
The Deutsche Nationalbibliothek lists this publication in the Deutsche Nationalbibliografie;
detailed bibliographic data are available on the Internet at http://dnb.dnb.de.

www.degruyter.com

Acknowledgments

The authors thank the European Association for Social Anthropologists (EASA) for funding the collaborative workshop between the Energy Anthropology Network and the Futures Anthropology Network in June 2019. We are grateful to the ENTPE, *l'école d'aménagement durable des territoires* and the *Laboratoire Aménagement Economie Transports* at the University of Lyon for hosting the workshop at which the ideas for this book were developed. We would like to acknowledge Charlotte Johnson, Oriane Manigault, Gabriela Cobana and Nicolas Nova who participated in the workshop for their contributions. Finally, we thank our editors at De Gruyter for their enthusiasm for this book and their support during its preparation.

https://doi.org/10.1515/9783110745641-001

Table of Contents

Sarah Pink, Nathalie Ortar, Karen Waltorp, and Simone Abram

1 Imagining energy futures: an introduction

Energy and futures are inextricably entangled in the narratives, hopes, and concerns of the second decade of the twenty-first century: hope for climate health is invested in an energy transition to renewable sources; the need to forecast energy demand better occupies electricity networks; engineers develop new technologies with the techno-solutionist promise of solving energy problems; governments, industry stakeholders and academic reviewers make choices about which energy-focused research agendas and projects to fund on the basis of their projected or expected future impact; electric car batteries are charged so that they will be ready for anticipated journeys; and everyday energy demand at home is often constituted through the mundane routines through which people in diverse circumstances accomplish what they need to do in order to be able to move forwards in life. In short, energy is bound up with the different stakeholders, scales and temporalities through which futures are imagined, desired, feared, planned or prepared for and anticipated, across diverse continents, countries, regions, neighbourhoods, communities and lives. During the period of writing this book we have also seen how demand for and access to energy has shifted during a global pandemic, climate events and war.

This means that to understand how and where energy is significant and meaningful to the lives of people, non-human species, and environments, we need to go beyond the conventional anthropological focus on the present immediately related to the past. Instead we need to engage with the question of how energy is entangled with futures. To do so is not only a step into a new temporality for energy anthropologies, it is also an inescapably critical step. It calls for us to ask how energy futures are being predicted, projected, and ostensibly shaped, by diverse stakeholders, and to consider our own role as anthropologists in proposing, and ultimately seeking to direct, different ways forward to answer challenges that have in recent scholarship often been associated with the concept of the Anthropocene.

Energy Futures addresses these circumstances. Born of a collaboration between energy anthropologists and futures anthropologists it brings anthropological expertise to insist on a rigorously analytical anthropology of energy, which is attentive to both the uncertainty and contingency of futures, and the creative capacity of people to improvise as they move forwards into circumstances about which their knowledge will always be incomplete.

https://doi.org/10.1515/9783110745641-002

In what follows we first critically review dominant visions for and of energy futures. We reveal how they are complicit with colonisation, capitalism, and narratives of technological solutionism and how they generally fail to understand, acknowledge, or attend to people as participants in energy futures. We then examine the conceptual foundations of a futures anthropology and energy anthropology approach to outline how they open up alternative ways of conceptualising energy futures. We discuss the methodological implications of this move towards future-focused energy anthropologies and analyses. We emphasise the need for change in the kinds and orientations of energy research funding. Finally we outline the book and its contributions.

Dominant visions of energy futures: forecasting and transition

Energy futures are not new – there has long been a need and intent to consider the future of energy production, demand and supply, and to understand how this intersects with the future of the planet and human lives, as well as government and corporate concerns. Formal energy futures have typically been quantified, existing in realms dominated by engineering, economics, and capital. Climate futures likewise have conventionally been quantified, by scientists, activists, and anyone who is trying to make a powerful point about environmental impacts, climate change, or otherwise, given the power of numbers to carry weight in political debate. The 3Ds –Decarbonisation, Decentralisation, and Digitalisation (Silvestre et al. 2018) – or 4Ds – Decarbonisation, Deregulation, Digitalisation, and Democratisation (DEEP 2018) – are expected, respectively, to be drivers in creating the green energy economy of the future and to transform the energy market. Yet energy and climate futures are also experiential domains; they can be understood and imagined through non-representational ways of knowing, including through deep indigenous knowledges of environment and country, in mundane moments of everyday life anywhere or during extreme weather and other climate events. We turn to these later; first we outline the dominant visions with which these are frequently in tension.

Global visions

Given the devastating impacts that global energy production is having on the earth's climate, changing energy systems is an urgent global concern. There are many and diverse interests at stake, and a correspondingly wide range of future imaginaries at play in thinking about energy futures. However, global political and investment strategies are normally based on quantified and technological assessments of potential solutions. Strategies for change, from those of United Nations (UN) Conferences of the Parties (COPs) to national and local agendas, most commonly call on a fairly narrow range of future projections supplied by agencies such as the World Energy Council (WEC) or International Energy Agency (IEA), or by commercial organisations such as BP. As Atle Midttun and Thomas Baumgartner pointed out some decades ago, although this kind of forecasting and modelling has been treated as technocratic and 'neutral', it has 'often been used for partisan purposes to push through, or oppose, certain energy developments' (1986: 219), leading to the clashes between industry-related and ecological forecasts that are so familiar today. The WEC[1] produces World Energy Scenarios, which they emphasise 'are designed to be used as a set to explore and navigate what might happen, not what should happen or what we want to happen'.[2] They offer three possible quantified scenarios, validated by consultation with their expert members based on surveys with leaders, which they recommend should be used to 'provide a useful, pre-decision thinking framework to help energy leaders engage with deep uncertainties and anticipate possible energy transition pathways'.[3] To achieve low carbon scenarios, the report declares, 'consumer behaviour needs changing'.[4] The IEA is an Organisation for Economic Co-operation and Development (OECD) agency governed by energy ministers or senior representatives from member countries, and meets every two years to determine its scope of work and budget. It is, in other words, explicitly a political organisation and although it professes to be fuel- and technology-ambivalent, it was founded in 1974 in response to the oil crisis.[5] Despite the consensus that models are by nature flawed, and despite the glaring possibility that industry projections could be self-serving, global predictions for continued and increasing demand for energy consumption continue to be rehearsed in all kinds of project meetings, and invoked as introductory stage-setting to inject urgency into problem-solving activities. The World Economic Forum (WEF) also

---

1 https://www.worldenergy.org/publications/entry/world-energy-scenarios-composing-energy-futures-to-2050.

2 https://www.worldenergy.org/transition-toolkit/world-energy-scenarios

3 https://www.worldenergy.org/transition-toolkit/world-energy-scenarios/long-term-world-energy-scenarios.

4 https://www.worldenergy.org/assets/downloads/World-Energy-Scenarios_Composing-energy-futures-to-2050_Executive-summary.pdf.

5 https://www.iea.org/about/mission, accessed 24 February 2021.

has a Global Future Council on Energy Transition,[6] which, it says, 'will provide a vision for the transition to a more sustainable, inclusive and affordable post-COVID world'. Its work is based on the assumption that technological solutions to our energy futures exist and should be deployed. Its report suggests that 'Like all technology-led transitions, the energy transition will unleash significant creative destruction, create large new opportunities for wealth formation, and will ultimately lead to greater prosperity and major societal benefits. Nevertheless, there will also be transition costs that need to be minimized and taken into account.'[7] No anthropologists appear in its list of 26 members.

If these views of energy futures continue to dominate, the world will be in much trouble. The evidence from anthropology is that the strategies of treating people as 'consumers' whose 'behaviour needs changing' (as the WEC does), or assuming that 'technology-led transitions' will lead to 'societal benefits' (as the WEF does), are based on unrealistic understandings of people and our relations with technology, and are therefore at great risk of failure. Rather, to work towards realistic, sustainable, ethical, and equitable futures, we need to acknowledge people in their diversity, their ways of knowing and their creativity. We believe these dominant approaches are misguided and problematic, but we do not wish to stop at critique. We want to open up a productive dialogue, and we seek to change a situation where dominant narratives are unable to account for the very issues that anthropological analysis demonstrates also need to be addressed. Anthropology complicates the narratives of global visions of energy futures; it generates new layers of uncertainty and proposes radical new ways forward. But, we argue, it needs to participate. The essential task is to generate a constructive anthropological voice in this global domain. This book is a step in this direction.

National forecasting and modelling

Modelling and forecasting is part of the way global visions are developed, but in the energy industry quantitative forecasting also plays a key role at the national level in seeking to determine how energy demand will take shape in the future and what infrastructure and other investments and policies are needed to respond to and influence this. The examples we discuss in this book demonstrate how this quantitative process involves multiple stakeholders from government

---

6 https://www.weforum.org/communities/gfc-on-energy-transition.
7 https://www3.weforum.org/docs/WEF_Energy_transition_known_and_unknown_2020.pdf.

and industry. In the UK, the National Grid produces annual Future Scenarios based on a two-by-two grid with axes such as 'speed of decarbonisation' versus 'level of decentralisation' (National Grid 2019). These use technical models such as UKTimes to project possible outcomes based on different rates of change. A closer look at National Grid scenarios shows these axes changing over time, with 2017 axes being 'prosperity' versus 'green ambition', while in 2020 and 2021 they adopted the axes: 'level of societal change/speed of decarbonisation', with a clear acknowledgement of the complexity of the challenge, 'not least because there are many ways to reach net zero, each with their own trade-offs.'[8] The scenarios claim to offer 'four different, credible pathways for the future of energy over the next 30 years'; whereas earlier scenarios included an option of 'business as usual', all the so-called pathways in the more recent scenarios move significantly towards decarbonisation, at different degrees of urgency. The possible options have changed not primarily due to changing technology but due to the growing acceptance across the energy industries that much of the responsibility to address climate change lies with them. Alongside this change comes an acknowledgement that such responsibility does not lie entirely within the scope of technological change or infrastructural improvement, but encompasses changes with a social basis. Yet, as argued above, there is a distinct lack of acknowledgement of a broad set of social science arguments, and a tendency to fall back on environmental psychology through which to create 'social acceptance'.

Yolande Strengers, Sarah Pink, and Larissa Nicholls (2019) have analysed existing approaches to energy forecasting, noting how similar logics inform these quantitative exercises across the UK, Australia and Europe: the UK National Grid Future Energy Scenarios claim that 'by providing a range of credible futures, we can be confident that the reality will be captured somewhere within that range' (National Grid 2016, cited in Strengers, Pink, and Nicholls 2021: 2); while the Australian Government's national science agency's (CSIRO)[9] major visioning project – the Future Grid Forum – proposed 'to inform and inspire a national conversation and provide a way forward for the sector, its stakeholders and, most importantly, all Australians' (CSIRO 2013, cited in Strengers, Pink, and Nicholls 2021: 14). The Australian Energy Market Operator (AEMO), manages Australia's electricity and gas systems and markets.[10] AEMO 'has produced electricity demand

8 National Grid FES2021: https://www.nationalgrid.com/uk/stories/journey-to-net-zero-stories/eso-future-energy-scenarios-next-30-years, accessed 15 January 2022.
9 http://www.csiro.au/en/About.
10 https://aemo.com.au/about/who-we-are.

forecasts for the National Electricity Market (NEM) since 2012'[11] and does so drawing on diverse data sources relevant to the task, including data relating to consumer trends, climate change and EV uptake[12]. However existing forecasting methods have not always predicted shifts in demand. For instance in the case of Australia the emergence of peak demand relating to air conditioning on extreme heat days led to electricity grid failures. These instances have demonstrated the need for deeper understandings of how human futures are likely to impact on the futures that forecasting modelling and scenario building seeks to predict (Strengers, Pink, and Nicholls 2019).[13]

Transition and decarbonisation

Other energy future projections are developed in the Transition Management field, with the aim to get to Net Zero emissions by particular end dates. Transition Management approaches seek to bring about step changes in energy systems and how people use energy. Yet they lack attention to the everyday worlds and the contingent circumstances that shape what people do with energy (Sharp et al. 2022).

The current energy transition to renewables involves a structural shift from fossil fuels such as oil, gas and coal to renewable energy sources such as solar, wind, and biogas and changing everyday life practices. It is debated in the arenas of academia, business, government, and industry, but often in ways that either disregard everyday worlds, or see them and the people who inhabit them as posing a behavioural challenge or barrier to energy efficient and sustainable futures and the technological solutions that are supposed to bring them about. The very choice of the term 'transition' downplays the turmoil and conflict caused by energetic uncertainty as transitionist imaginaries suggest a gentle, gradual, consensual change and tend to depoliticize its real implications (Loloum, Abram, and Ortar 2021). As anthropologists have shown, renewable energies have also been part of the long succession of abuses and exploitation of human and natural resources carried out by foreign actors with the complicity of local actors

---

11 https://aemo.com.au/en/energy-systems/electricity/national-electricity-market-nem/nem-forecasting-and-planning/forecasting-and-planning-data/nem-electricity-demand-forecasts

12 https://aemo.com.au/-/media/files/electricity/nem/planning_and_forecasting/nem_esoo/electricity-demand-forecasting-methodology-information-paper.pdf?la=en

13 The Digital Energy Futures project led by Yolande Strengers and Sarah Pink in Australia (discussed in Ch. 2) has subsequently worked with forecasters to develop new forecasting methods which will accommodate ethnographic futures insights.

(Howe et al. 2015), discussed in Ch. 5. To convert solar energy into electricity depends on the same extractive industries, globalised production networks, and electronic waste streams that characterised the industrial exploitation of resources, land, and labour (Cross 2019). Wind or solar parks have social effects on the regions that host them and are part of unequal power relations, both north–south and north–north (Knight and Argenti 2015).

Transition management, moreover, when envisaged as individual responsibilities to deliver visions of desired futures, assumes that it is possible to be granted 'social licence' from people (classified as consumers, users, publics, or citizens) to implement change, to which they will subsequently adapt their everyday lives. There have recently been attempts to meaningfully engage the concept of social licence – a term most commonly used in the context of the extractive industries – into discussions about energy transition and supply- and demand-side management. The term was engaged by the Australian Council of Learned Academies' 2020 Australian Energy Transition Research Plan,[14] and in a 2021 international Technology Collaboration Programme (TCP) report, *Social License to Automate*[15]. For Sophie Adams and colleagues, SLA [social licence to automate] 'provides a framework to understand the (mis)alignments between the expectations of actors within the energy system on the one hand, and household practices, sense of control and stake in the energy system, on the other' and reveals the negotiations that occur in each domain (Adams et al. 2021: 2). It is debatable whether engaging terminologies and concepts which originate in the extractive industries is the right position for anthropologists. As presented by Adams and colleagues, it takes a step forward informed by the social sciences to offer a more nuanced and realistic approach, in terms of creating dialogue and engagement in ways recognisable to industry and government stakeholders. Yet such alignments simultaneously risk diluting arguments for a more radical paradigm shift in how everyday worlds and automated energy systems might be co-constituted and exacerbating existing power imbalances.

Both energy forecasting and Transition Management approaches project into the future, the former working forward towards particular years, and the latter working backwards from a desired outcome. However, the reality is that precisely how everyday energy futures will come about is impossible to know, and without acknowledging and seeking to work with the messiness and contingency of the everyday such approaches are difficult to apply.

---

14 https://acola.org/wp-content/uploads/2021/06/acola-2021-australian-energy-transition-plan. pdf#page=31
15 https://userstcp.org/wp-content/uploads/2021/10/Social-License-to-Automate-October-2021. pdf

Automation, digital capitalism, and digital colonialism

As climate change and the consequences of global heating intensify, transnational and local public institutions and industries continue to place their bets on digitalisation as a technical solution to the problem of carbon emissions (Szeman and Boyer 2017). Industries proclaim a 'fourth industrial revolution' based on AI and cloud computing (Schwab 2016) and likewise industry and consultancy reports assume that emerging digital and automated technologies will be solutions to problems of energy demand and supply and will assist in the transition to net zero emissions (Dahlgren et al. 2020), discussed in Ch. 2. Energy companies turn to the technology industries to find 'smart' solutions to the intermittency of renewable energy and to the increased pressures put on local and national grids by the ongoing electrification and digitalisation of transport and industrial production, regarding digital technologies as tools to secure grid stability (Blondeel et al. 2021: 11).

Emerging automated systems and devices offer technological possibilities, on the basis of which it is possible for organisations to build other technologically driven future imaginaries, thus making infrastructures that not only support the delivery of energy management systems and change processes, but can be interpreted as 'anticipatory infrastructures' upon which other future imaginaries are built (Pink et al. 2022, and Chs. 4 and 5 below). In the case of energy futures, these enable imaginaries of automated technologies that will manage energy demand, such as automated wireless charging systems, and energy demand management systems that take control of energy supply to household devices out of the hands of householders.

The context of the growing and possible further automation of the energy industry is envisaged as bringing many benefits. Yet in doing so, like the ambitions towards energy transition discussed earlier, its limitation is that stakeholders in this field have tended to engage with people as consumers, whose behaviour may be changed through the application of technological solutions, once they trust and accept them. These logics are typical across industry and government approaches to emerging technologies (Pink 2022a), and are entrenched in the dominant technologically solutionist narratives of contemporary society. Moreover, this world where technologies are increasingly automated and connected itself requires excessive amounts of ambient energy, which in turn requires new digital infrastructure (Bresnihan and Brodie 2020). This includes the immense energy demand of data storage facilities, discussed in Ch. 4, as well as the further energy costs associated with the resource extraction required for the production, transportation, installation, and maintenance of these technologies. The lack of attention both to people and to the 'invisible' costs of automated and connected technologies in energy futures relates to the wider issue of how imagined, envisaged, and planned energy futures are both embedded in and could exacerbate existing inequalities and colonial continuations, on which Chapter 5 elaborates, both within countries and in terms of how their costs are unequally distributed across the world.

One aspect of this is the growth of what has been called *digital capitalism*, whereby 'the accumulation of capital, decision-power and reputation are mediated by and organised with the help of digital technologies and where economic, political and cultural processes result in digital goods and digital structures' (Fuchs 2022: 28). Digital capitalism is often discussed in dystopian narratives, warning of what could happen if its logics were to play out, often through existing examples which highlight how inequalities have come about through unethical uses of automated and connected technologies. For example, building on approaches that study platforms as infrastructural intermediaries in everyday life and positioning digital platforms within the political economy/ecology literature on rent(ierism), Jathan Sadowski (2020: 13) critically defines platforms as a dominant form of rentier in contemporary capitalism through data extraction, digital enclosure, and capital convergence (van Dijck and Poell 2015). Sadowski argues that platforms have latched onto and inserted themselves into production, circulation, or consumption process. By doing so platforms capture rents from all economic activity they mediate and as such should be considered as an evolution and expansion of rentier capitalism. The *digital inequalities* associated with digital capitalism are manifold. The production of big data as a commodity involves processes best understood as accumulation through dispossession (Harvey 2004): as technology users enter into tacit data licence agreements with the firms that create and control the technology, they are dispossessed of the right to control those data. This could provide the preconditions for a new stage of capitalism that can barely be imagined, but for which the appropriation of human life through data would be central (Thatcher, O'Sullivan, and Mahmoudi 2016). Government data have also imposed a new regime of surveillance, profiling, punishment, containment, exclusion, and worsening inequality (Eubanks 2018). Inequalities are also woven throughout 'the digital inequality stack in many ways from differentiated access, use, and consumption, literacies and skills, and production' (Robinson et al. 2020).

These possible generic effects of digital technologies are coupled with the exclusionary nature of digital networks, which are built and imagined in such a way as to exclude or include certain people and places, as discussed in Chs. 4 and 5. Copper and fibre optic cables constitute the Internet, showing how the invisible 'cloud' connecting devices and data is rather a material set of cables, servers and antennas, with implications for politics, histories and imaginaries (Hu 2015, Furlong 2021, Maguire, Watts, and Winthereik. 2021, Starosielski 2015, Thylstrup 2019). Nicole Starosielski (2015) highlights that offshore cable connections are often built through – rather than to – coastal and rural communities that host secure cable landing and access points, making them outposts for tech and telecoms companies operating and delivering services elsewhere. Moreover, through its very infrastructures, the resources needed, and its access, the cloud is at the origin of what has been called *data colonialism*, which combines the predatory extractive practices of historical colonialism with the abstract quantification methods of computing (Couldry and Mejias 2019) creating new forms of inequalities between the global North and South. To store data and have them ready to hand, data centres have repurposed the industrial and military infrastructures of the twentieth century, connecting 'our digital present to our industrial past' (Pickren 2017: 22, Furlong 2021, see also Ch. 4). It is not only the 'natural' cooling of northern climates which are making northern countries aspiring hubs of the new digital economy (Brodie 2020, Johnson 2019, Vonderau 2018, Velkova 2021) but also their connections to major cable networks, which were laid across colonial routes of oceanic trade and communication (Furlong 2021). As argued by Nick Couldry and Ulises Mejias (2019), digital colonialism's future incidence is still unknown 'as the scale of this transformation means that it is premature to map the forms of capitalism that will emerge from it on a global scale'.

These possible digital and technological energy futures are frightening, and it is urgent for us to attend to them. The existing literature reveals their possibilities and their logics, often focusing on a likely progression to dystopian ends. They warn us of what could happen. However, these do not need to represent the realities of our energy futures. In fact the dystopian narratives of critical social theorists frequently fail to account for the capacity of people to disengage, not to participate in the automated futures they portray. The ethnographic evidence is that people do not necessarily sign up for the journey towards dystopian digital futures (Pink et al. 2022), and specifically that they often do not envisage themselves relinquishing control of their data, energy systems and access to energy companies or other organisations (Pink 2022b).

The evolving relationship between emerging digital and automated technologies and our local and global energy futures needs considerably more work. There is an important role for anthropologists who can be sensitive to how change happens, how power relations are enacted, how inequalities are experienced, and what these questions mean for intervention. It is to the role of anthropology in our energy futures that we turn our attention in the next section.

Thinking energy futures anthropologically

The need to attend to energy futures anthropologically, in observing how technical approaches tend to close the field of possibilities, has long been part of anthropologies of energy. In the 1980s, Laura Nader (1981) pointed out the inconsistencies of a system 'founded on the deleterious fear of change in lifestyles' despite 'the very wide range of lifestyle choices available in any plausible energy future' (Nader 2010: 241). So how can energy futures be envisaged? How can current systems and infrastructures be turned around? How can whole countries across the globe move from carbon states to decarbonised states? Who will imagine the road from C to D? And what knowledge will they rely on to start that journey?

Science studies tell us how crucial the imagination is in generating particular futures, and the actions and policies that flow from it. Janet Stewart shows how the visual language of fossil technologies colonises the everyday imagination from which futures are conjured (Stewart 2016), making extraordinary global claims appear to be little more than common sense through a teleological narrative. Theorists such as John Law have shown how models and metaphors have been implicated in the shaping of modern states (1986), with data emerging early on as an essential governmental technology for acting at a distance, to govern distant territories and trades. A number of sociologists and historians have shown how numbers, statistics, calculability and probabilities have supported the science of the state – political science – in shaping political agendas. Peter Miller and Nikolas Rose showed how double entry book keeping came to define the budgetary systems of modern states, defining which elements could be quantified and therefore recognised by the state (1990), while Ian Hacking (1986) showed how these enumerations led to the 'making up of people' who would reflect the categories defined by calculable quantities – observations developed in Evelyn Ruppert and Stephan Scheel's study of how central data practices are to the making of a people at the European level (2021).

These socio-technical principles apply equally to energy futures, shaped as they are by the technological imaginaries of their designers; the engineers and mathematicians who model current and postulated energy systems, defining as they go along the limits to possibility, and embedding cultural expectations and judgements into the evaluation of technological designs. Hence it is important to look at how new technologies are brought into practice, and how infrastructures are shaped and controlled. This task speaks directly to current concerns in social anthropology of addressing hybrid-social worlds, that is, socio-material, socio-technical, and beyond-human articulations. Examining how technologies are imagined, crafted and adopted becomes an anthropological lens through which to unpack the terminologies and concepts that support and are supported by technological imaginaries. What are the actions at a distance performed by electrical grids, for example? What are the limits to possibility of the notion of 'energy' itself? (see Abram, Winthereik, and Yarrow 2019, Daggett 2019, Anand, Gupta, and Appel 2018).

Research into energy transitions has shown how the technical systems in place lead to inertia and not only limit possible action in the future (Nader 1981, Araújo 2014) but frame a specific thinking that prevents the ability to 'shift gears' (Nader 2004). For Mette High and Jessica Smith, the alleged inevitability of the energy transition 'hamper[s] our ability to engage and respond to it' (2019: 11) further by narrowing the universe of possibilities. As noted above, existing methods of scenario planning and forecasting used in the energy sector often tend to prioritise a limited number of possible futures (Strengers et al. 2019). High and Smith (2019: 11) argue that 'the overall framework of energy transitions has narrowed the scope of how anthropologists understand and engage in ethical dilemmas posed by energy'. Understanding these dilemmas therefore presents new challenges for anthropology to include the questions, desires and concerns of diverse but interdependent groups of people.

Such thinking, which accounts for the future, is now present in the majority of anthropological energy research, whether it relates to the exploitation of oil (Weszkalnys 2016, Witte 2018), new fossil resources (Szolucha 2018), or renewable energies (Boyer and Howe 2019, Knight 2012, 2018), or to the imagined futures of smart grids and the automation of residential energy supply (Pink 2022b). Energy resources and infrastructure – existing or planned – generate 'citizen hopes, desires and aspirations' (Weszkalnys 2016: 161) that 'saturate the conceptions of time and the future' (Ferry 2016: 185). This growing emphasis on futures in energy anthropology has coincided with a significant expansion of interest in the anthropology of the future, and futures anthropologies (Collins 2020). This creates new opportunities to engage with energy futures through anthropology, and indeed to understand all anthropological futures research as inevitably having something to say about possible energy futures, given the inevitability of energy across the sites of all anthropological investigation.

However, the turn to futures in anthropology is not one unified move, rather it comes from and stands for divergent theoretical threads and conceptual categories. This means that the approach to futures in anthropology which is followed has analytical consequences, and subsequently informs both the nature of the stories that we are able to tell from our ethnographies and the interventions we are able to make. That is not to say necessarily that one approach is better than the other, but that we should carefully consider which we take in any given set of circumstances, because theory has effects. In academia the theoretical approach we take will mean that our work is accepted by, or drawn into dialogue with, certain camps within the discipline. In interdisciplinary collaborations the concepts we engage may open or close particular shared interdisciplinary categories, debates, and ways of knowing. Moreover, in engaged research, theory can limit or enable the kinds of interventions we make.

Chapter 2 suggests how such differences might play out. There we compare two anthropological approaches to futures: Rebecca Bryant and Daniel Knight's (2019) anthropology of the future on the one hand, and on the other Sarah Pink and Juan F. Salazar's (2017) futures anthropology. These are two very different ways of looking at futures in anthropology, to the extent that they are not in dialogue with each other in these two works. Bryant and Knight's work is a study of anticipation in everyday life in the present, based on the social practice theory (SPT) of Theodore Schatzki. Pink and Salazar, in contrast, argue for a radical departure from the anthropology of the present, towards a futures anthropology that roots itself in the contingent nature of present and futures. The relevance of this approach for everyday life is that, coming together with design anthropology, it builds on the growing emphasis in anthropology on indeterminacy, uncertainty (Akama, Pink, and Sumartojo 2018), the imagination (Sneath, Holbraad, and Pedersen 2009), emergence (Smith and Otto 2016), and a more generally processual (rather than culturalist) underpinning to anthropological thought (Ingold 2000). In Ch. 2 Sarah Pink and co-authors engage both design-anthropological and SPT approaches, demonstrating how approaches informed by SPT produce opportunities to trace futures in the present, while futures anthropology approaches, with their leaning towards design anthropology, enable more interventional approaches that seek to engage with futures as modes of generative uncertainty, as much as to study them.

Chapter 3 engages directly with the governance of energy futures through the technological imaginaries and the hopes and fears they generate or confront. Thinking with the generation of energy and the resources and commitments that go into changing energy systems and regimes, the chapter addresses a shift in anthropological focus from studies of anticipation to the contradictory horizons that different futures appear to offer (cf. Guyer 2007). It considers how different kinds of energy futures are both envisaged and contested in relation to different moments that shape so-called energy transitions, showing how divergent understandings can become positioned as 'resistance' or 'conflict', while some are considered apolitical. It shows how closely energy futures are entangled with technologies of government, bringing the Science and Technology Studies (STS) critique of governance into dialogue with discussions about energy infrastructures and recent research on extractivism.

Chapter 4 engages with futures through digitalisation and the planning of data infrastructures and consumption. It examines how their futures are planned and to what effect, and attends to the everyday implications of these infrastructures on the future of neighbouring communities as well as the future ecologies they are producing. Data centres, it reveals, are paradoxical infrastructure as they are 'both solid and durable and evaporative and itinerant; it is built and grown, rigid and fluid, meant to last but doomed to be outmoded, ruined, and exceeded' (Howe et al. 2015). As such, data centres enabled and accelerated many of the effects associated with the Anthropocene, and yet might be the best way to ameliorate the contemporary predicament. To engage with their future theoretically, the authors engage with the anticipatory temporalities that drive the data centre industry (Hu 2015, Taylor 2021, 2022) and the implication of 'preparedness', a notion loosely defined by Frédéric Keck (2016) as 'a state of vigilance cultivated through the imagination of disaster'. Through a focus on preparedness anthropologists have revealed the logics of how dystopian future scenarios produced by practitioners whereby threatening events – from pandemic outbreaks to cyberattacks to extreme weather events – are constructed as inevitable and unpreventable, but potentially manageable (Taylor 2021).

Data centres thus participate in an 'economy of anticipation' (Cross 2015) in which socio-technical imaginaries are used to divert attention from the present (Abram and Weszkalnys 2013) and build local and national accounts of a data-driven desirable future. These socio-technical imaginaries participate in a 'solutionist' (Morozov 2013) innovation paradigm, whereby the futures they imagine often propound technological solutions to human and environmental problems, assuming that neither people nor the environment have agency in such processes. Julia Velkova (2021: 665) defines these dynamics, and the economic and social relations that they install, as a thermopolitics which supports conservative agendas. Such agendas – even when presented as disruptive – are likely to be appropriated by powerful stakeholders in strategies that maintain the status quo and develop digital capitalism. In Ch. 4 Nathalie Ortar and co-authors (dis)entangle these narratives through ethnography to shed light on how the various stakeholders build their legitimacies through an economy of anticipation and by doing so try to close down alternatives futures.

In Ch. 5, the future-anthropologies agenda (Salazar, Pink, Irving, and Sjöberg 2017) introduced in Ch. 2 is extended further in a move to unsettle a solutionist paradigm and pay attention to other ways of relating to landscapes, resources, and energies than the 'standing reserve' (Heidegger 1977, Waltorp, Lanzeni, Pink and Smith 2022) logic of extraction for growth without limit. Karen Waltorp has engaged with how various flows of images across physical, digital, and imaginal realms figure in people's lives, relations, and futures, in ways that mean having your phone charged is of utmost importance (Waltorp 2017, 2020, 2021). In Chapter 5, Waltorp and co-authors look at what it takes to enable such a flow of images in a digitalised world; and show that water is primary across field sites in Chile and Botswana. Water is needed to produce energy, it is needed to cool down cables, and as we see in Ch. 4 the data centres that store our data. At the same time digital infrastructures are also enabling alliances of circulating images in mainstream and social media of indigenous people fighting for their approach to land and resources. In northern Norway and in Southern Africa there is no wish not to have this infrastructure and access, yet the tech giants entering the region are energy-hungry. The consequence for the electricity system is increased pressure in a situation of recurrent scheduled blackouts. Chapter 5 further engages with research on extractivism that Ch. 3 introduces elaborating on regional colonial histories, asking what continuations exist in terms of whose techno-social imaginaries and futures come to count and carry more weight in terms of what is deemed preferable, probable and even possible. Chapter 5 argues for engaging with decolonial scholarship, activism, and alliances to open up more possible energy futures that have people and environment at the centre, and challenge the sedimented logic of treating the earth and other beings as standing reserves.

Methodologies for energy futures

An interventional anthropology for energy futures needs new methodologies, and a new methodological openness to the approaches of other disciplines and stakeholders. If we are to have a chance to influence the course of energy futures we need to be able to collaborate with disciplines and organisations in this field, both through our own research strengths and by applying our anthropological sensitivity to understand other ways of being and seeing to our own in the contexts of collaboration and the other actors involved.

As implied throughout this introduction, both immersive long-term ethnography and a range of increasingly experimental applications of ethnography – represented across the chapters of this book – form an important basis of response to the dominant narratives, practices, and assumptions which we wish to complicate and intervene in. Ethnographic practice foregrounds new narratives, other possible futures, and stories that are frequently invisible in the agendas towards transition to net zero and the assumptions that emerging automated and connected technologies will solve societal and global energy problems. It involves sharing aspects of peoples' lives with them in the present, and as they experience possible futures in order to comprehend the sensory, affective, and expressive modes through which energy futures coalesce, how these configure with the global, local, and everyday circumstances of life, and how these circumstances configure and reconfigure over time.

Anthropological ethnography must also take seriously the everyday ethics (Mattingly and Throop 2018) both of the people among whom and situations in which we do fieldwork and of ethnographic practice. As anthropologists working in the energy futures field we have an ethical responsibility to our participants, in the conventional sense of fieldwork ethics. But don't we also have a responsibility to go beyond accounting for how energy is implicated in the circumstances of their lives and futures, and to also engage what we learn *with* them towards energy futures that are just and fair? Anthropological ethnography can show how energy futures are situated in relation to the multiplicity of things, other species, processes, and possibilities that people and energy are co-implicated with. It thus enables us to bring to the surface these modes of connectedness and relationality. Energy futures emerge not in a vacuum but in worlds already full of politics, power, and taken-for-granted notions of what works and how. Ethnography makes these relations visible and their implications obvious, both as a general principle and in a move that pulls into view the very differentiated cultural taken-for-granted ways that these relations, their implications, and the ways they are experienced play out in different places and across different scales. These relations, visible through ethnography, play out different connections and logics to the causal chains of top-down transition focused on technological solutions that dominate in the narratives outlined earlier in this introduction. However, to ensure that we can make these differences and their implications clear, we need methodologies for engagement.

Therefore, to confront energy futures we need to be methodologically bold as outlined in the future anthropologies network's (FAN) collectively developed manifesto (in Salazar et al. 2017: 1–2). The first three points identify futures anthropologists as critical and 'engaged with confronting and intervening in the challenges of contested and controversial futures'; collaborating without fear of contamination or incapacity from exposure to other disciplines; and decentring the human to embrace 'larger ecologies and technological entanglements'. Each of these points resonates strongly with the methodologies we need to engage for an energy futures anthropology, and is played out in the following chapters as we show how our work contests the dominant narratives outlined above, engages with other disciplines and stakeholders, and accounts for environments, other species, and technologies as participants with humans in emerging energy futures.

The futures anthropology manifesto is also strongly interventional. Futures anthropologists 'get our hands dirty' and 'we are ethical, political and interventionist and take responsibility for interventions' (in Salazar 2017 et al.: 2). Methods for making energy anthropologies public and interventional are emerging across both examples discussed in this book and more broadly amongst anthropologists of energy. They include: new modes of reporting ethnographic energy futures in ways that dialogue with industry reporting (e.g. Dahlgren et al. 2020, Strengers, Pink, and Nicholls 2021, Nicholls et al. 2021); new techniques of visualising possible futures and therefore making the issues themselves visible to diverse groups of stakeholders (Dahlgren et al. 2021, and Ch. 2); new models of writing and co-authorship across projects and places (Waltorp and ARTlife Film Collective 2021), such as that developed in this book, and by James Maguire and colleagues in 2021; creating new modes of storytelling about energy futures accessible across academic and other sectors, including documentary filmmaking, such as *Digital Energy Futures* (Pink 2022b); making interactive websites (DigiSAtproject.com), tracing encounters between logics and feral ecologies growing out of – but beyond the control of – human-built infrastructures as in the 'Feral atlas' of the more-than-human Anthropocene (Tsing et al. 2021); creating new materials for intervention, such as design cards (see Pink et al. 2017); hackathons (Flipo et al. 2022); future speculative design scenarios of colour packaging for waste reduction, theatre with janitors and inhabitants in social housing areas, and creating waste robots in collaborations between students, industry stakeholders, and local inhabitants as was done in the course and book project 'Question Waste: Experiments between ethnography and design' (Waltorp and Halse 2013).

Energy futures anthropology methodologies are currently developing in the light of the need for new approaches to researching possible futures ethnographically, interdisciplinarily, and as part of stakeholder-partnered projects, driven by the impulse for intervention. This creates a dynamic field of activity which we expect to develop further as energy futures anthropology gains ground.

What's next?

In energy-related research and innovation the vast majority of funding has gone to the natural and technical sciences (Overland and Sovacool 2020). Efforts towards interdisciplinarity have had limited effect so far (Baum and Bartkowski, 2020), although the European Commission has maintained its commitment both to mainstreaming Social Sciences and Humanities (SSH) across all its funded research and to creating opportunities for dedicated SSH-led research where needed.

There is clear evidence of social science and humanities narratives being excluded from energy funding calls and government consultations in Europe (Royston and Foulds 2021), and we hope to spark a broader conversation around how these excluded disciplines, including anthropology, can contribute to energy policy and research. For example, Ch. 2 illustrates how well-designed stories, like comic strips depicting industry imaginaries about e-cars and everyday life, create an environment in which to explore knowledge in a situated, context-dependent way (Mourik, Sonetti, and Robison 2021). We argue that storytelling can form the foundation for inclusive and participatory future thinking and visioning and echo the Future Anthropologies Manifesto (in Salazar et al. 2017: 1–2) calling for 'a politics of listening attuned to a diversity of voices and [to] tell stories that are imaginative, illustrative and informative.' This book demonstrates the important steps that can be taken when anthropological research into energy futures is funded. These opportunities have not only enabled us to contribute knowledge and practice but also to develop and propose the vision for energy anthropology futures outlined in this introduction. However, we argue that approaches to funding need to change; anthropology needs to be recognised in energy research as a discipline able to understand and integrate other disciplines and stakeholders (see von Wirth et al. 2020, Ryghaug et al. 2021, Foulds et al. 2020, Flipo et al. 2020 for relevant questions).

About *Energy Futures*, the book

This book was written collectively by a team of 24 authors spread across the world, during a global pandemic. It is structured into four key chapters, as outlined above, each of which was curated by one of the book's editors but co-authored with our collaborators. Each chapter addresses key issues and arguments relating to its respective area of focus. Importantly each chapter also pivots its discussion around a set of *ethnographic cases* each authored by the contributors to the book and based on their own fieldwork. The chapters can be read in any order or as single pieces of work.

References

Abram, S., and Weszkalnys, G. (2013). *Elusive promises: planning in the contemporary world*. New York: Berghahn Books.
Abram, S., Winthereik, B. R., and Yarrow, T. (eds.) (2019). *Electrifying anthropology: exploring electrical practices and infrastructures*. London: Bloomsbury Academic.

Adams, S., Kuch, D., Diamond, L., Fröhlich, P., Henriksen, I. M., Katzeff, C., Ryghaug, M., and Yilmaz, S. (2021). Social license to automate: a critical review of emerging approaches to electricity demand management. *Energy research and social science*, 80, 102210. https://doi.org/10.1016/j.erss.2021.102210

Akama, Y., Pink, S., and Sumartojo, S. (2018). *Uncertainty and Possibility*. London: Bloomsbury.

Anand, N., Gupta, A., and Appel, H. (2018). *The promise of infrastructure*. Durham, NC, and London: Duke University Press.

Araújo, K. (2014). The emerging field of energy transitions: progress, challenges, and opportunities. *Energy research and social science*, 1, 112 – 121.

Baum, C. M., and Bartkowski, B. (2020). It's not all about funding: fostering interdisciplinary collaborations in sustainability research from a European perspective. *Energy research and social science*, 70, 101723.

Blondeel, M., Bradshaw, M. J., Bridge, G., and Kuzemko, C. (2021). The geopolitics of energy system transformation: a review. *Geography compass*, 15(7), 1–22. https://doi.org/10.1111/gec3.12580.

Boyer, D., and Howe, C. (2019). *Wind and power in the Anthropocene*. Durham, NC: Duke University Press.

Bresnihan, P., and Brodie, P. (2020). New extractive frontiers in Ireland and the Moebius strip of wind/data. *Environment and planning E: nature and space*, 4(4), 1645 – 1664. https://doi.org/10.1177/2514848620970121.

Brodie, P. (2020). Climate extraction and supply chains of data. *Media, culture and society*, 42(7 – 8), 1095 – 1114. https://doi.org/10.1177/0163443720904601

Bryant, R., and Knight, D. (2019). *The anthropology of the future*. Cambridge: Cambridge University Press.

Collins, S. G. (2008). *All tomorrows cultures: anthropological engagements with the future*. New York: Berghahn.

Couldry, N., and Mejias, U. A. (2019). Data colonialism: rethinking Big Data's relation to the contemporary subject. *Television and new media*, 20(4), 336 – 349. https://doi.org/10.1177/1527476418796632

Cross, J. (2015). The economy of anticipation: hope, infrastructure, and economic zones in South India. *Comparative studies of South Asia, Africa and the Middle East*, 35(3), 424 – 437.

Cross, J. (2019). The solar good: energy ethics in poor markets. *Journal of the Royal Anthropological Institute*, 25(S1), 47 – 66. https://doi.org/10.1111/1467-9655.13014

CSIRO (2013). *Change and choice: the future grid forum's analysis of Australia's potential electricity pathways to 2050*. Newcastle, NSW.

Daggett, C. N. (2019). *The birth of energy: fossil fuels, thermodynamics and the politics of work*. Durham, NC: Duke University Press.

Dahlgren, K., Pink, S., Strengers, Y., Nicholls, L., and Sadowski, J. (2020). *Digital energy futures: review of industry trends, visions and scenarios for the home*. Emerging Technologies Research Lab. Monash University. https://www.monash.edu/__data/assets/pdf_file/0008/2242754/Digital-Energy-Futures-Report.pdf.

Dahlgren, K., Pink, S, Strengers, Y, Nicholls, L, and Sadowski, J. (2021). Personalization and the smart home: questioning techno-hedonist imaginaries. *Convergence*, 27(5), 1155–1169. doi:10.1177/13548565211036801.

DEEP (2018, November 18). The 4Ds transforming the energy market. *Medium.* https://me
   dium.com/@d33p/the-4ds-transforming-the-energy-market-1fb61fba385e.
Eubanks, V. (2018). *Automating inequality: how high-tech tools profile, police, and punish the
   poor.* London: Picador, St Martin's Press.
Ferry, E. (2016). Claiming futures. *Journal of the Royal Anthropological Institute*, 22(S1),
   181–188.
Flipo, A., Sallustio, M., Ortar, N., and Senil, N. (2021). Sustainable mobility and the
   institutional lock-in: the example of rural France. *Sustainability*, 13(4). https://doi.org/10.
   3390/su13042189
Flipo, A., Sallustio, M., Senil, N., Mao, P., and Ortar, N. (2022). *Expériences de
   recherche-action avec les partenaires.* Paris: ADEME.
Foulds, C., et al. (2020). *100 social sciences and humanities priority research questions for
   energy efficiency in Horizon Europe.* Cambridge: Energy-SHIFTS.
Fuchs, C. (2022). *Digital capitalism: media, communication and society, volume three.*
   London: Routledge.
Furlong, K. (2021). Geographies of infrastructure II: concrete, cloud and layered (in)visibilities.
   Progress in Human Geography, 45(1), 190–198. https://doi.org/10.1177/
   0309132520923098
Guyer, J. (2007). Prophecy and the near future: thoughts on macroeconomic, evangelical and
   punctuated time. *American ethnologist*, 34(3), 409–421.
Hacking, I. (1986). Making up people. In T. C. Heller, M. Sosna, and D. E. Wellbery (eds.),
   Reconstructing individualism: autonomy, individuality and the self in Western thought.
   Stanford, CA: Stanford University Press. 222–236.
Harvey, D. (2004). The 'new' imperialism: accumulation by dispossession. *Socialist register*,
   40, 64–87.
Heidegger, M. (1977). *The question concerning technology and other essays.* New York:
   Harper and Row.
High, M. M., and Smith J. M. (2019). Introduction: the ethical constitution of energy
   dilemmas. *Journal of the Royal Anthropological Institute*, 25, 9–28.
Holmes, T., Lord, C., Ellsworth-Krebs, K. (2021). Locking-down instituted practices:
   understanding sustainability in the context of 'domestic' consumption in the remaking.
   Journal of consumer culture. doi:10.1177/14695405211039616.
Howe, C., et al. (2015). Paradoxical Infrastructures. Ruins, retrofit, and risk. *Science,
   technology and human values.* http://doi.org/10.1177/0162243915620017.
Hu, T.-H. (2015). *A prehistory of the cloud.* Cambridge, MA: MIT Press.
Ingold, T. (2000). *The Perception of the Environment.* Abingdon: Routledge.
Johnson, A. (2019). Data centers as infrastructural inbetweens. *American ethnologist*, 46,
   75–88.
Keck, F. (2016). 'Preparedness.' theorizing the contemporary. *Fieldsights*, September 30.
   https://culanth.org/fieldsights/preparedness
Knight, D., and Argenti, N. (2015). Sun, wind and the rebirth of extractive economies:
   renewable energy investment and metanarratives of the crisis in Greece. *Journal of the
   Royal Anthropological Institute*, 4(21). http://doi/10.1111/1467-9655.12287
Law, J. (1986). On the methods of long distance control: vessels, navigation and the
   Portuguese rute to India. In Law, J. (ed.), *Power, action and belief: a new sociology of
   knowledge?* Routledge and Kegan Paul. 234–263.

Loloum, T., Abram, S., and Ortar, N. (2021). Introduction: politicizing energy anthropology. In T. Loloum, S. Abram, and N. Ortar (eds.), *Ethnographies of power: a political anthropology of energy.* New York: Berghahn. 1–23.

Maguire, J., Watts L., and Winthereik, B. R. (eds.) (2021). *Energy worlds in experiment.* Manchester: Mattering Press.

Midttun, A., and Baumgartner, T. (1986). Negotiating energy futures: the politics of energy forecasting. *Energy policy,* 14(3), 219–241.

Miller, P., and Rose, N. (1990). Governing economic life. *Economy and society,* 19, 1–31.

Mourik, R. M., Sonetti, G., and Robison, R. A. (2021). The same old story–or not? How storytelling can support inclusive local energy policy. *Energy research and social science,* 73, 101940.

Nader, L. (1980). *Energy choices in a democratic society.* Washington, DC: National Academy of Sciences.

Nader, L. (1981). Barriers to thinking new about energy. *Physics today,* 34(9), 99–104.

Nader, L. (2004). The harder path – shifting gears. *Anthropological quarterly,* 77(4), 771–791.

Nader, L. (ed.) (2010). *The energy reader.* Chichester: Wiley-Blackwell.

National Grid (2016). *National Grid stakeholder feedback document 2016: future energy scenarios.*

http://fes.nationalgrid.com/media/1156/fes-2016-stakeholder-feedback-document.pdf

Overland, I., and Sovacool, B. K. (2020). The misallocation of climate research funding. *Energy research and social science,* 62, 101349.

Pickren, G. (2017). The factories of the past are turning into the data centers of the future. *Imaginations: journal of cross-cultural image studies,* 8(2), 22–29.

Pink, S. (2022a). *Emerging technologies: life at the edge of the future.* London: Routledge.

Pink, S. (2022b). *Digital energy futures.* Documentary film. Emerging Technologies Research Lab, Monash University, Victoria, Australia.

Pink, S., and Salazar, J. F. (2017). Anthropologies and futures: setting the agenda. In J. Salazar, S. Pink, A. Irving, and J. Sjöberg (eds.), *Anthropologies and futures.* London: Bloomsbury.

Pink, S., Lanzeni, D., and Horst, H. (2018). Data anxieties: finding trust in everyday digital mess. *Big data and society.* https://doi.org/10.1177/2053951718756685.

Pink, S., Sadowski, J., and Nicholls, L. (2022). Digital technology and energy imaginaries of future home life: Comic-strip scenarios as a method to disrupt energy industry futures. *Energy research and social science,* 84, 102366.

Pink, S., Berg, M., Lupton, D., and Ruckenstein, M. (eds.) (2022). *Everyday automation: experiencing and anticipating emerging technologies.* Abingdon: Routledge.

Pink, S., Leder Mackley, K., Morosanu, R., Mitchell, V., and Bhamra, T. (2017). *Making homes: ethnographies and designs.* London: Bloomsbury.

Robinson, L., et al. (2020). Digital inequalities 3.0: Emergent inequalities in the information age. *First Monday,* 25(7). https://doi.org/10.5210/fm.v25i7.10844

Robison, R., et al. (2020). *100 social sciences and humanities priority research questions for smart consumption in Horizon Europe.* Cambridge: Energy-SHIFTS.

Ruppert, E., and Scheel, S. (2021). *Data practices: making up a European people.* London: Goldsmiths Press.

Ryghaug, M., et al. (2020). *100 social sciences and humanities priority research questions for transport and mobility in Horizon Europe.* Cambridge: Energy-SHIFTS.

Sadowski, J. (2020). The Internet of landlords: digital platforms and new mechanisms of rentier capitalism. *Antipode*, 52(2), 562–580.

Salazar, J., Pink, S., Irving, A., and Sjöberg, J. (eds.). *Anthropologies and futures: researching emerging and uncertain worlds*. London: Bloomsbury.

Schwab, K. (2016). *The fourth industrial revolution*. World Economic Forum: London: Portfolio Penguin.

Sharp, D., Pink, S., Raven, R., and Farrelly, M. (2022). Transition management and design anthropology: an interdisciplinary agenda for decarbonising cities. In K. Araújo (ed.), *Routledge handbook of energy transitions*. London: Routledge.

Smith, R. C., and Otto, T. (2016). Cultures of the Future: Emergence and Intervention in Design Anthropology. In R. C. Smith, K. T. Vangkilde, M. G. Kjærsgaard, T. Otto, J. Halse, and T. Binder (eds.), *Design anthropological futures*. London: Bloomsbury Academic. 19–36.

Sneath, D., Holbraad M., and Pedersen, M. A. (2009). Technologies of the Imagination: An Introduction. *Ethnos: Journal of Anthropology*, 74(1), 5–30.

Silvestre, M. L. D., Favuzza, S., Sanseverino, E. R., and Zizzo, G. (2018). How decarbonization, digitalization and decentralization are changing key power infrastructures. *Renewable and sustainable energy reviews*, 93, 483–498. https://doi.org/10.1016/j.rser.2018.05.068

Starosielski, N. (2015). *The undersea network*. Durham, NC: Duke University Press.

Stewart, J. (2016). Visual culture studies and cultural sociology: extractive seeing. In *The Sage handbook of cultural sociology*. Sage: Newbury Park, CA. 332–334.

Strengers, Y., Pink, S., and Nicholls, L. (2019). Smart energy futures and social practice imaginaries: forecasting scenarios for pet care in Australian homes. *Energy research and social science*, 48, 108–115.

Strengers Y, Dahlgren, K., Nicholls, L., Pink, S., and Martin R. (2021). *Digital energy futures: future home life*. Emerging Technologies Research Lab (Monash University). Melbourne, Australia. https://www.monash.edu/__data/assets/pdf_file/0011/2617157/DEF-Future-Home-Life-Full-Report.pdf.

Szeman, I., and Boyer, D. (eds.). (2017). Energy humanities: an anthology. Baltimore, ND: Johns Hopkins University Press.

Taylor, A. R. E. (2021). Future-proof: bunkered data centres and the selling of ultra-secure cloud storage. Journal of the Royal Anthropological Institute, 26(S1), 76–94.

Taylor, A. R. E. (2022). Bunkers, data, preparedness: from the mushroom cloud to the computing cloud. New media and society.

Thatcher, J., O'Sullivan, D., and Mahmoudi, D. (2016). Data colonialism through accumulation by dispossession: new metaphors for daily data. Environment and planning D: society and space, 34(6), 990–1006. doi: 10.1177/0263775816633195

Thylstrup, N. B. (2019). The politics of mass digitization. Cambridge, MA: MIT Press.

Tsing, A., Deger, J., Saxena, A., and Zhou, F. (2021). Feral atlas: the more-than-human Anthropocene. Stanford, CA: Stanford University Press.

van Dijck, J., and Poell, T. (2015). Social media and the transformation of public space. Social media + society, July. doi:10.1177/2056305115622482

Velkova, J. (2021). Thermopolitics of data: cloud infrastructures and energy futures. Cultural studies, 35(4–5), 663–683. https://doi.org/10.1080/09502386.2021.1895243

Vonderau, A. (2018). Technologies of imagination: locating the cloud in Swedens global North. *Imaginations: journal of cross cultural image studies*, 8(2), 8–21.

von Wirth, T., et al. (2020). *100 social sciences and humanities priority research questions for renewable energy in Horizon Europe*. Cambridge: Energy-SHIFTS.

Waltorp, K., Lanzeni, D., Pink, S. and Smith, R. C. (2022). Introduction: An Anthropology of Futures and Technologies. *An Anthropology of Futures and Technologies*. London and New York: Routledge.

Waltorp, K. (2017). Digital technologies, dreams and disconcertment in anthropological worldmaking. In J. Salazar, S. Pink, A. Irving, and J. Sjöberg (eds.), *Anthropologies and futures: researching emerging and uncertain worlds*. London: Bloomsbury. 117–133.

Waltorp, K. (2020). *Why Muslim women and smartphones: mirror images*. London: Routledge.

Waltorp, K. (2021). Multimodal sorting: the flow of images across social media and anthropological analysis. In A. Ballestero and B. R. Winthereik (eds.), *Experimenting with ethnography*. Durham, NC: Duke University Press. 133–150.

Waltorp, K., and ARTlife Film Collective (2021). Isomorphic articulations: notes from collaborative film-work in an Afghan–Danish film collective. In L. Di Puppo, L. Martínez, and M. D. Frederiksen, *Peripheral methodologies: unlearning, not-knowing and ethnographic limits* (Anthropological Studies of Creativity and Perception). London: Routledge. 115–130.

Waltorp, K., and Halse, J. (2013). Introduction. In J. Dresner, J. Halse, V. Johansson, R. S. Troelsen, and K. Waltorp (eds.), *Question waste – experimental tactics between ethnography and design*. Copenhagen: The Royal Danish Academy of Fine Arts, School of Design.

Weszkalnys, G. (2016). A doubtful hope: resource affect in a future oil economy. *Journal of the Royal Anthropological Institute*, 22(S1), 127–146.

Witte, A. (2018). *An uncertain future—anticipating oil in Uganda*. Göttingen: Göttingen Series in Social and Cultural Anthropology.

Sarah Pink, Katherine Ellsworth-Krebs, Michiel Köhne,
Elisabet Dueholm Rasch, Nathalie Ortar, Aurore Flipo,
and Kari Dahlgren

2 Everyday futures, spaces, and mobilities

Introduction

The everyday is a dynamic site through which to investigate energy futures. It is where the mundane but crucial activities, feelings, and relationships that underpin more spectacular or visible domains of life play out. Expressed directly for academics, the everyday is where, to be dressed to deliver an in-person lecture, you need to have done your laundry. The everyday is composed of many such background activities, materialities, technologies, and sensory experiences, all demanding energy for their ongoing maintenance, and ultimately entangled with the climate, political, and health crises, resource extraction, inequalities, infrastructures, and technologies which characterise our present and possible future environments.

In anthropology, design and science and technology studies (STS) there is growing attention to mundane anticipation (Pink and Postill 2019, Bryant and Knight 2019) and to how smart technologies are implicated in the anticipatory modes of home (Johnson 2020, Michael 2016, Strengers and Kennedy 2020, Knox 2021), mobility (Pink, Fors, and Glöss 2018) and work (Pink, Ferguson, and Kelly 2022). In this chapter we understand everyday energy futures from three perspectives.

First we stress that energy and technology use are configured in the everyday narratives of life in homes, mobilities and work in the present (Strengers 2013, Pink and Leder Mackley 2015, Pink, Fors, and Glöss 2019, Velkova et al. 2022), and acknowledge the many inequalities of digital capitalism (e.g. Eubanks 2018, Sadowski 2020). We need to foreground tensions between the messy realities of the everyday with the visions of energy and technology futures presented by politically and economically powerful stakeholders; how do messy and contingent everyday life realities complicate the sanitized, quantified mainstream future visions of energy demand proposed by consultancies, industry, and technology news media? Second, we emphasise the need to examine how energy futures are creatively imagined in everyday life. How does energy use participate in people's everyday values, hopes, expectations, and anxieties for their near and far futures? Third, how does energy use come about through these messy reali-

https://doi.org/10.1515/9783110745641-003

ties of life in the present and as it is imagined for the future, as always relational to other everyday priorities and activity?

To engage with these questions ethnographically we foreground the stories of people in affluent nations whose access to everyday energy is predominantly via electricity grids, fuel stations, and consumer products, and shaped by the techno-solutionist agenda of neoliberal governments and powerful industry stakeholders. In pursuing their stories we raise a set of issues. A richness of ethnographic material has been produced with participants living in such circumstances, enabling comparison. This is ironically indicative of how research funding has supported energy research initiatives in wealthy countries, and signals the need for a more diverse agenda (see Chs. 4 and 5). Moreover, dominant narratives about energy and technology futures proposed by consultancies, industry, and policy bodies appear usually to refer to such societies. Here we seek to complicate their future visions from the very sites that they superficially appear to be consistent with.

In Australia and Europe, the sites we write from, we have experienced recent climate events – including bushfires and extreme heat in Australia, flooding in the United Kingdom, and the COVID-19 pandemic. Such events are both the outcomes of resource depletion and have energy demand implications. Across all our fieldwork and writing sites, lockdowns and home-working between lockdowns have focused work and life on the home, increasing the use of digital technologies in homes, and generating speculation on the reduced energy demands of digital mobility and reduced commutes (Holmes et al. 2021). This chapter creates a close-up encounter with 'anthropology at home' (Amit 2000), yet in doing so dislodges the claims of conventional anthropologies at home. Doing research about energy *futures* in the everyday, or *everyday energy futures,* disrupts the possibility that anthropologists are at home. Although we inhabit our futures in our visions of them, the future is in fact the home of no one.

This, we argue, is precisely why we need to respond; we need to occupy energy futures with *possible* futures, of the kind that complicate the predicted futures that are so often advanced by dominant narratives. Everyday energy futures are already being colonised by the consultancies, the energy and technology industries, and by governments. Often their moves are well-meaning, they are also often paternalistic and usually support corporate capitalism. Such organisations predict and claim futures through their practices of envisaging 'better' more sustainable futures, automated for the supposed convenience and comfort of 'consumers' and to optimise their effectiveness. They colonise futures and seek to bring people into line with their visions, in supposing people will enact different everyday lives, routines, and priorities once they 'properly' use future technologies, systems and modes of automation. We need to make everyday energy futures our new intellectual and applied anthropological home.

Anthropologies of everyday energy and futures

Anthropologies of energy in everyday life vary vastly in their localities and sites, their modes of ethnographic practice, engagements with institutional, corporate, government, other-sector, or activist stakeholders, and their sources of funding. Compared with other disciplines, anthropologies of everyday energy in homes took off slowly, initially in Scandinavian countries with a stronger tradition of applied research in this field (Henning 2005, Wilhite 2005, Bille 2018). There are different anthropological modes of working with everyday futures, all of which have been fruitfully engaged in energy anthropologies.

One trend is the *anthropology of the future* (Bryant and Knight 2019), which uses social practice theory (SPT) to study how the present is shaped by anticipated futures. Sociologists such as Elizabeth Shove in the UK, Kirsten Gram Hannsen in Denmark, and Yolande Strengers (2013) in Australia developed SPT as a lens for rethinking energy demand in homes, with a critical agenda against the behaviour-change policies advanced by neoliberal governments that tended to frame everyday life in the home as culpable for energy waste. Scholars have engaged SPT to argue both that everyday life practices shape energy demand, and energy use is not a behaviour that can be changed by appealing to rational actors. Ellsworth-Krebs (this chapter) applies SPT to discuss how the socio-materiality of the everyday shapes energy demand in the UK and Australia. Pink's collaboration with Strengers has combined SPT with futures anthropology (Strengers, Pink, and Nicholls 2019). Social practices have been studied in order to understand how historically situated practices performed today can tell us something about opportunities for change. SPT has also been applied to understanding how changes might be imagined and experimented with in the future and practice-centered design for change initiatives is a growing field of research. The example that Michiel Köhne and Elisabet Dueholm Rasch elaborate suggests that everyday practices 'may have limited ability to shape sociotechnical imaginaries themselves' but are an important factor in the energy transition (Schelhas 2018: 186); they are also limited in their use for designing possible futures (Pink and Leder Mackley 2015), but important for understanding everyday practice in the present.

The other approach to futures in anthropology, *futures anthropology* (Pink and Salazar 2017, Pink 2021), is rooted in phenomenological and design anthropological theory and investigates possible futures by working in speculative or experimental modes. It is particularly useful for attending to how energy, emerging technologies, and digital data intersect in the everyday present and possible futures. In increasingly datafied environments (Couldry and Mejias 2019), big-data analysis, predictive analytics and emerging technologies are integral to visions of how future home energy demand will be created and mitigated (Strengers et al. 2021) and technological solutionist visions see EVs as pivoted to become ubiquitous, using secure blockchain transactions to pay for automatically wireless charging (Pink 2022, Ortar and Ryghaug 2019). Design anthropological approaches define the everyday as a site of ongoing emergence (Smith and Otto 2016, Akama, Pink, and Sumartojo 2018), where forms of resistance, creative adaptation, and invention characterise people's evolving relationships with and modes of learning with technology. Combined with the futures anthropology emphasis on the contingency of everyday life in the present and future, such work alerts us to the dynamic nature of the everyday, and the impossibility of holding it still for prediction, as dominant narratives seek to. Design anthropological approaches to emerging technologies (e.g. Pink et al. 2022) emphasise the need to respond to such narratives by attending to how these technological possibilities will play out in the real everyday. Speculative approaches, typically associated with design and 'the experiment', an STS methodology, have increasingly become engaged with this question in anthropological and sociological work on everyday energy. For instance, sociologist Mike Michael's collaboration with designer William Gaver produced a speculative object placed in participants' homes in the United Kingdom (Michael 2016); Julia Velkova and colleagues undertook research alongside the trial of an automated energy demand management system in Sweden (Velkova, Magnusson, and Rohracher 2022); and anthropologist Hannah Knox's experimental work with participants and their own hand-made energy monitoring data leads her to argue for a new propositional approach (Dányi et al. 2021: 84).

These speculative studies exceed the conventions of long-term ethnographic fieldwork by creating generative experiments in the present. They additionally generate insight concerning how people live and learn with speculative technologies, systems, services, or processes that do not usually inhabit their everyday present, when they are accommodated *into* it. Whether or not we should call scenarios like technology tests and trials, or people's experiences of them, possible futures is debatable. But they do create situations where people's ongoingly emergent futures are opened to new possibilities, which could not have occurred without research, design, or experimental interventions. Such interventions offer the only empirical knowledge we have of how people experience and engage with new energy-related technologies in the everyday. They provide unique possibilities through which to question or complicate the assumptions about what everyday human futures will look like with new energy technologies promoted in dominant narratives.

Another way to engage with everyday energy futures interventionally is through the notion of energy democracy, which emphasises the importance of efforts by citizens to exercise more control over energy decisions and as such to construct their desired (renewable) energy future. Rather than looking for a technological fix, energy democracy questions who controls energy, to what end, and to whose benefit (Fairchild and Weinrub 2017). Energy democracy scholars propose decentralised ways of governing energy (Burke and Stephens 2017, 2018), but focus on people who can afford to invest in renewable energy solutions. Energy democracy is thus not only an element of an energy future, but also a way of giving form to and moving towards that renewable energy future. The example discussed in the case presented by Köhne and Rasch below adds a new dimension to the discussion on energy democracy by introducing how social housing residents can be part of the construction of energy democracy and an inclusive renewable energy future.

However, in advocating for energy futures from the everyday, we do not directly pitch the everyday against industry, government, policy, and such like, or see them as operating at different poles of a continuum. Rather we see benefits in bringing together diverse stakeholders in everyday energy futures. We collectively believe and show how our work should be relevant to a wide range of organisations including: governments, municipalities or social housing corporations aiming to reach net zero emissions targets or engage tenants in renewable energy programmes; energy companies to guide energy demand forecasting and future infrastructural investments; property developers in reprioritising away from a sales-focused model to consider changing demographic trends and housing needs; designers, architects, and planners; energy entrepreneurs; local energy cooperatives; and coworking organisations. We must enter into dialogue with such organisations, to demonstrate the benefits of everyday anthropological thinking. We moreover need to engage collaboratively with other disciplines to understand the possibilities for researching and intervening in visions of plausible everyday energy futures. The theoretical tools available to our field outlined above likewise should not be activated to compete with each other, but to provide the agility through which to achieve the engagement required for our participation.

Future visions and everyday energy futures

Industry and consultancy visions frequently predict future everyday life through techno-solutionist narratives. These tend to be based on assumptions about the impact technologies will have on people (usually referred to as users or consumers) when – or *if* – they use them as intended. In particular, the need to decarbonise the energy system raises a number of key challenges for the energy sector, in which people's everyday actions and decisions are often perceived as 'barriers' to an optimised energy system. It is subsequently assumed that such behaviours can be overcome either through economistic visions which suggest consumers can be influenced through the right mix of incentives and price signals, or through erasing their inefficient behaviour entirely through outsourcing energy decision making to automated systems (Sadowski and Levenda 2020). However, such visions ignore the complex ways energy use is entangled with everyday practices, priorities, and ethics. People are often visioned into such technologically determinist views of energy futures as personas, whose rational actions will align with the ambitions of efficient energy systems. Social scientists have critically responded to this caricaturing of human experience and practice, by showing up the personas for what they are. Yolande Strengers's (2013) Resource Man, Charlotte Johnson's (2020) Flexibility Woman, and Kari Dahlgren and colleagues' techno-hedonist (2021) personas outline the dissonance between what we consider to be realistic possible everyday futures, and the kinds of people that are imagined to inhabit everyday energy futures.

Visions of future everyday energy demand often have utopian feel-good narratives, precisely because they focus on possible adjustments to sustainable living and energy demand reduction, without accounting for the reality that the contingencies of everyday life will lead to resource depletion in other invisible ways. By situating our research in the everyday we can view the tensions between utopian visions of energy demand reduction at the local scale and the energy and resource depletion that new modes of automation and technologies demand globally, and/or in other global sites. Coworking reveals that the very technologies that make the reduced travel and local focus of coworking possible, generate e-waste, deplete minerals, and themselves demand energy for their production, maintenance, transportation, and data storage. Electric vehicles generate environmental costs through their production and shipping. While, due to health and social inequality issues, questions of overcrowding have dominated discourses on floor- area trends, issues of excess and over-consumption have been missed. Social housing tenants experience a very different sense of ownership over energy decisions from what is represented in utopian feel good-narratives where renewable energy is owned and produced by more privileged groups to maintain comfortable lifestyles.

The next four sections bring this to life through ethnographies of the everyday.

The cases

In the following four sections, we discuss examples from across Australia and Europe concerning: the tensions between industry visions and everyday imaginaries of future electric vehicle (EV) charging (Kari Dahlgren and Sarah Pink); how energy demand is produced at the intersection between mobility, home, and work futures, through a focus on the rise of coworking (Nathalie Ortar and Aurore Flipo); a reconsideration of dominant visions of the relationship between everyday priorities relating to house size and energy demand (Katherine Ellsworth-Krebs); and a consideration of ownership as a central pillar of energy democracy (Michiel Köhne and Elisabet Dueholm Rasch). In doing so we collectively bust the techno-solutionist myths that are part of assumptions that: coworking leads to a sustainable energy future; people inevitably wish for bigger homes in the future, which impacts on energy efficiency; tenants have ownership over energy decisions about their homes, disregarding how this may be reduced through limited financial resources or of a different kind while using frugality as a strategy rather than renewable investment; e-mobility can be generalised without changing everyday practices. We advocate for energy futures reimagined from the everyday.

Complicating smart charging electric vehicle futures

Kari Dahlgren and Sarah Pink

In Australia the climate crisis has in the last years manifested visibly in bushfires, flooding, and extreme heat and these environmental and weather events have direct consequences for domestic energy demand. Within this, electric vehicles (EVs) are frequently seen as part of a solution or techno-fix to the problem of reducing carbon emissions, which requires the buy-in of both future drivers and government in providing infrastructure and incentives. Our Digital Energy Futures (DEF) project has explored how people anticipate or imagine how EVs could figure in their possible everyday futures, by allowing participants to contemplate how the complexity of their own lives, values, and desires for different futures complicate mainstream industry and policy narratives. Our research reveals the potential of ethnographic futures research to contest techno-solutionist narratives of the energy sector.

Our work in DEF is in collaboration with sociologists, Yolande Strengers, Larissa Nicholls, and Rex Martin. DEF is Australian Research Council and industry-partner funded, with two Australian energy distribution companies, AusNet and Ausgrid, and Energy Consumers Australia, and in its final stages collaborates with quantitative energy forecasters (Strengers et al. 2021). Its first and second stages, discussed here, involved a review of industry reports, developing probe materials, and online ethnography designed to invoke and compare industry-framed futures and possible everyday futures.

The probes we created were a series of comic strips informed by the qualitative content analysis (Schreier 2012) of 64 digital technology and energy sector reports (international but focused on trends likely to affect Australian households), which identified current industry trends, predictions, and visions for how everyday practices are anticipated to change in the near (2025–2030) and medium-far (2030–2050) futures. Reading these reports also entailed an immersion into the logics, discourses, and future visions of industry and policy, making them a fieldsite of future imaginations, speculations, and predictions from which we gathered key claims and imaginaries of the energy and technology industry, and consultancies (Dahlgren et al. 2021). In order to challenge these dominant visions of the future we needed to articulate them, and to engage participants with them. The key findings of the review were published in a report (Dahlgren et al. 2020) and synthesised into six comic strip scenarios, which playfully depict how the digital technology and energy trends and visions are predicted to impact everyday life. As aggregated versions of the future trends and visions found in the reports reviewed, the scenarios do not represent the research team's own future visions, but those that have been extrapolated from the review of reports. They distilled the discourse of these reports into an accessible and entertaining form of encountering their implications for the future home, which we used in our ethnography. Below we discuss one comic strip scenario, representing visions of electric vehicles and their integration into the 'smart grid' and energy demand management through the smart home (Fig. 2.1).

Dominant EV narratives

Electric Vehicles (EVs) are expected to overtake combustion engine vehicles in the coming decades, predicted to reach 100 % of all new vehicles sold in Australia by 2040 (KPMG 2018a). Our analysis showed EVs are often viewed as a crucial technology for decarbonising society, allowing the replacement of petroleum with electricity, ideally derived from renewable sources. EVs promise a simple solution to petroleum powered transport without having to significantly alter travel patterns or expectations of individual mobility. However, they raise another problem. For the electricity sector, electric vehicles represent a significant increase in electricity demand, which if not properly managed, could potentially stretch energy infrastructure to the breaking point. As the title of a KPMG report on EVs asks, "Is the Energy Sector Ready?"

Unlike oil, electricity is a peculiar commodity that resists storage. Supply and demand must be carefully matched to avoid outages, but both fluctuate radically based on weather conditions (for renewables) and how humans pattern their social life. Australia's electricity sector is affected by 'peaky' conditions, where on hot summer afternoons electricity demand can double as Australians collectively switch on their air conditioners. The thought that someday they will also all be plugging in their electric vehicles to charge when they arrive at their air conditioned homes is a frightening prospect to those charged with securing the electricity supply, and would undermine potential decarbonisation gains, should increased fossil-fuelled electricity be required to meet such demand.

However, our analysis revealed that even in the face of this new problem, the logics of technological solutionism are unrelenting, as a new solution is proposed in the form of "smart charging", which also brings into focus the broader goal for energy demand management issues to be solved by the 'smart grid'. The 'smart grid' promises a responsive electricity system that manages supply from an increasing number of distributed energy resources (DER) such as rooftop solar PV, and controls demand through load-shifting digital technologies that are responsive to dynamic pricing. Consultants to the energy sector warn that "if business models, market design and technology do not align consumer incentives with efficient behaviour, even a modest increase in electric vehicles could strain our generation and network infrastructure" (Deloitte 2018). This emphasis on "consumer incentives with efficient behaviour" represents what Strengers (2013) calls the Resource Man vision of energy consumers – a tech-savvy and energy-interested version of homo-economicus – which she points out is central to the sector's vision of a price-responsive energy system. However, beyond market design, Deloitte's words emphasise technology's envisioned role in making behaviour more 'efficient', in two interrelated ways. First, EVs can be integrated into the electricity grid, becoming mobile batteries providing network services to distributed energy resources. Second, Artificial intelligence (AI) and Automated Decision-Making (ADM) can ensure that the use of EVs and their charging are optimised in response to electricity availability.

Thus, according to these narratives, EVs can be transformed from a burden into a potential solution to other technology-created solutions-turned-problems, that is, the need to integrate the increasingly distributed energy system, particularly household solar PV, into stable, reliable, and balanced loads. One particular manifestation of this is vehicle-to-home, or vehicle-to-grid technologies, whereby the electric vehicle serves as a mobile battery, and its charge and discharge are utilised to level electricity loads while maintaining sufficient charge for travelling. This technology requires particular patterns of use, high levels of flexibility, price responsiveness, and sufficient charging infrastructure. This vision also privileges a particular form of vehicle user: Australia's National Science Agency CSIRO portrays 'a household that has access to charging via both home off-street parking and at their normal place of daytime parking (i.e. at work or in a carpark)' (Graham and Havas 2019). These future visions assume that applied as such, technology directly improves people's lives. CSIRO states that for 'Over nearly a century, we've been improving the lives of people everywhere with our science' (CSIRO nd).

ADM is increasingly viewed as a solution to a range of assumed 'barriers to adoption' of EVs, including customers' range anxiety and the perceived inconvenience of charging (Pink 2022). From the perspective of the energy sector, it is clear that ADM is also crucial to their vision of how the electricity system will manage to incorporate high rates of EV ownership and turn EVs from a potential threat to the electricity system to a solution to the increasing prevalence of distributed energy generation. Electric vehicles and ADM for energy demand management may represent opportunities for increased decarbonisation, but they reflect a technological-solutionist framing of climate change. They offer easy technological solutions that don't require any fundamental alterations of our practices or social structures. As Morozov has described for techno-solutions, 'In promising almost immediate and much cheaper results, they can easily undermine support for more ambitious, more intellectually stimulating, but also more demanding reform projects' (Morozov 2013: 9). EVs don't require us to question our mobility practices or question the market logics of continued economic growth, planned obsolescence, and green consumerism, nor any of the interconnected environmental impacts of our Anthropocenic lives, such as the multiple manifestations of planetary damage, or modernity's ontological dualisms which ignore human and nature mutualities and entanglements. Thus the decarbonisation they promise involves shifting rather than solving problems.

Further, technological solutions often fail when they meet up against social life, which is always made in complex assemblages of humans and non-humans. Therefore ethnography presents an opportunity to challenge these future visions through people's everyday lives, values, and desires for their futures. However this also entails a challenge for ethnography in terms of how to engage participants in research in the sites of possible futures which exist beyond their immediate experience and imaginations (Markham 2021, Dahlgren et al. 2022).

Comic-strip scenarios for EV futures

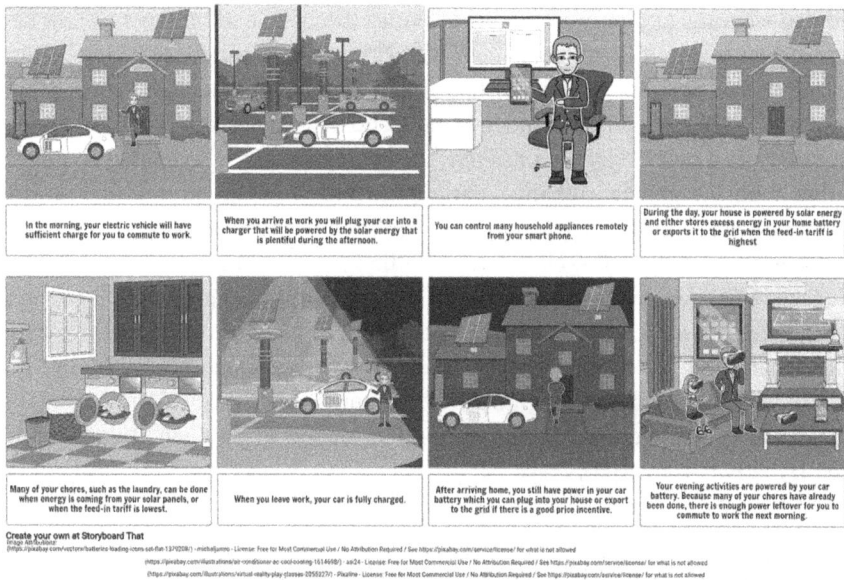

In the morning, your electric vehicle will have sufficient charge for you to commute to work.

When you arrive at work you will plug your car into a charger that will be powered by the solar energy that is plentiful during the afternoon.

You can control many household appliances remotely from your smart phone.

During the day, your house is powered by solar energy and either stores excess energy in your home battery or exports it to the grid when the feed-in tariff is highest

Many of your chores, such as the laundry, can be done when energy is coming from your solar panels, or when the feed-in tariff is lowest.

When you leave work, your car is fully charged.

After arriving home, you still have power in your car battery which you can plug into your house or export to the grid if there is a good price incentive.

Your evening activities are powered by your car battery. Because many of your chores have already been done, there is enough power leftover for you to commute to work the next morning.

Fig. 2.1: The Active Smart Charging Commuter comic strip. Published with permission of the Digital Energy Futures project.

The comic strip in Fig. 2.1 represents the ideal version of the smart charging EV owner with vehicle-to-grid technology portrayed in the reports analysed. It shows the availability of charging at home and workplace, flexible use of smart household appliances in response to price incentives, and both the smart home appliance users and the 'set and forget' ADM that automatically trades energy in relation to dynamic pricing. This vision of the EV future is the idealised techno-solutionist response to the need to decarbonise society. Practices including the commute, washing and drying clothes, and high evening energy use are maintained, but technology has seamlessly enabled this ongoing pattern of consumption and mobility while promising to deliver the energy required through optimised renewable sources. Household chores are flexible, capable of being performed in optimal energy load conditions, whereas the commute is regular, routine and predictable following an archetypal middle-class pattern.

The comic strip reveals problematic assumptions underlying these visions of the future: that householders have a high interest in energy usage data and price signals, and a willingness and ability to make changes in response to these data and incentives, which past research indicates is not always the case (Strengers 2013, Hargreaves, Wilson, and Hauxwell-Baldwin 2018). The reports assume that most people's energy usage is flexible, while research indicates that everyday practices can be difficult to shift through data and price signals or are considered essential and non-negotiable activities (Nicholls and Strengers 2015). Reports privilege an affluent 'consumer' who can afford to purchase technologies, an individual operator or head of household who controls technologies, rather than how such decisions are made in households that contain a mix of adults, children, and pets. They also assume the repetitiveness of everyday routines, rather than the constant change, variations, and ongoing exceptions that are part of everyday life.

The comic strips allowed a form of ethnographic access to these industry envisaged futures, by the very people they are being imagined for: Australian households. Research participants were able to think through how their own life did and did not fit into such a future. The comic strips were used after in-depth interviews and ethnographic home tours with participants, most of which were conducted online due to the interruption of the Covid-19 pandemic (see Dahlgren et al. 2022).

Complicating the narratives in the everyday

We focus on one example, of Cindy, who had recently completed a home renovation which saw her install many efficiency upgrades into her home in a regional Australian town. Cindy was particularly proud of her solar panels, and derived pleasure from using the mobile app that allowed her to check in real-time her household electricity usage, how much was coming from the grid, and how much she was selling back to the electricity grid. When purchasing her solar panels, she intentionally picked ones that could be connected to a battery but was waiting for batteries to become less expensive before investing in one. She was currently unemployed and spent much of her time tending to her garden and caring for her ageing mother who lived with her. Her adult son also lived with her but was busy working in construction.

Our ethnographic household visit involved an extended tour of her house, where she explained and recreated her everyday routines, and household practices, such as cooking, cleaning, and laundry. This visit was one of our first, just before the Covid-19 pandemic disrupted in-person fieldwork. Many of her practices were motivated by a strong conservation ethic. For example, she was particularly adamant that she would never purchase a clothes dryer, screaming 'No!' when we asked her. She found it excessively wasteful. Hanging laundry on the outside line worked perfectly well. She kept a bucket under the taps to save the water that she ran before it was hot enough for a shower or washing dishes, for watering the garden. However, she was concerned about the comfort of her house, and particularly concerned with keeping the house a comfortable temperature for her mother, so ran the air conditioning and heat regularly. She recognised this was in tension with her conservation priorities, but her care duties were more important to her.

After the home tour, we sat down with Cindy and showed her the comic strip scenario in Fig. 2.1. After briefly explaining the scenario, she shook her head in agreement, uttering 'makes sense'. We asked if she thought this scenario would work for her, she thought for a moment and said that there would need to be some checks in place to ensure she still had enough charge for her car. Using the scenario after the household visit meant we were already familiar with Cindy's household practices and routines, and could draw on this to ask how her life would fit into the scenario. When we asked if she would be willing to relinquish some of the comfort derived from heating or cooling at her preferred level, in order to preserve the battery, she said that she was not willing to sacrifice her comfort. She told us she was 'not money motivated', but added, 'for other people' it might work.

The scenario also shows an automated laundry machine operating when solar energy is available. Initially Cindy thought she could use this, but then paused and asked, 'but who would hang the laundry out?' Thinking some more she solved the problem herself by imagining it as a washer/dryer combo. When we asked her if she could imagine herself using a dryer in such a future, even more adamantly than previously, she fully shouted "No!, but for other people! I've gone as far as I can putting myself into this scenario."

This distinction Cindy drew between her own life, and how others might live was consistent across our research participants. Technology-led futures might 'make sense' in the abstract, but once imagined in the realities of the participant's own everyday life, they begin to fall apart, making more problems than they solve. Presenting future scenarios to people in the context of their home lives revealed both layers – both the ways that their logics convince, and how people's actual lives complicate the narratives.

We asked Cindy if the same situation would make sense for her son, who works in construction and goes to different worksites each day. She explained how it would not since

> if you're going to work at one place this is very relevant, but I don't know how relevant this scenario is for most people these days because a lot of people do work from home, a lot of people...this is a very old school scenario this 'go to one place of work.' In 2050 is this going to be the scenario that people work in, doubt it

What did she think the future would be like? She continued:

> A lot of people working from home and being at home 24/7; people travel so much for work, so no one is leaving their car in one spot; a lot of people are *in* their car for work. A lot of people are working online, or like [my son] a lot of people are traveling to different places for work...This scenario is taking an old scenario and putting it into the future which may not translate.

Cindy recognised that not only did the future envisioned in the scenario not fit into the conditions of her own life, but also that it was inherently conservative: it privileged middle-class routines, of commuting to a single place of work with dedicated charging infrastructure, rather than the 'flexible', contract, and gig work that characterises much of the Australian workforce's existence. This was only further disrupted through the experience of the Covid-19 pandemic.

Calling for ethnographic energy futures

The reports envisioned flexible energy usage for domestic appliances and tasks, but expected predictable patterns of mobility. In contrast Cindy, while willing to do her laundry at a different time, could not be flexible since she needed to be physically present to hang out her laundry and needed to keep the house at a comfortable temperature for her mother. Although she had access to her energy data through the smartphone app which tracked the solar production and her electricity consumption linked to her solar PV system, these data did not translate into flexible practices around energy consumption.

As Cindy rhetorically asked of the comic strip: 'Who lives like this?'

It may be that it is the authors of the reports we reviewed: those working in the technology and energy sector, consultants, and policy advisors, who might themeslves be versions of Resource Man (Strengers 2013). However, the future they are envisioning, planning for, and thus contributing to making is not the future that many Australians see themselves in. The techno-solutionist logic that has infiltrated this vision of the EV attempts to write humans out of the future. This technologically deterministic view of social life, which paints humans as the problem itself, underlies visions of EV futures. This is not to discount the potential benefits of electrifying transport, but if we are to address the multi-dimensional and complex problems that the Anthropocene presents, we need a broader lens. We need to recognise that without a broader mandate for the energy transition that takes into account everyday life and priorities, we end up only shifting problems rather than solving them.

Coworking space as low energy futures?

Nathalie Ortar and Aurore Flipo

Coworking spaces are seen as one of the outcomes of the 'second digital revolution'. They have gradually established themselves as the heralds of new work expectations. In parallel, development of new information and communication technologies (ICTs) has made possible new lifestyles and generated new forms of remote work (Ortar 2018, Sajous 2019) as well as new needs for meeting and copresence (Benedetto-Meyer and Klein 2017, Flipo and Ortar 2020). Consistent with a 'Californian spirit' that combines technology and ecology, community practices and market economy, coworking spaces rely on the assumption that the transition towards a more sustainable future cannot be achieved without the digital transition (Monnoyer-Smith 2017), while their energy costs are not evaluated. Composed of people with neither hierarchical nor customer-to-supplier relationships, coworking spaces are based on the idea of serendipity and randomness. They have certain characteristics typical of start-up culture, including forms of scenography (creative rooms, paperboards) and furniture typically associated with the home rather than work (sofas, hammocks, table football, etc.).

In France, coworking spaces have expanded thanks to the support of public authorities and private finance. In 2019, with over 1,200 coworking spaces it had more than most countries on the planet (Leducq 2021). While some closed as a result of the spring 2019 first COVID-19 lockdown, with the dramatic growth in home-working many new coworking places have opened since then, despite the ongoing sanitary restrictions (Leducq 2021). As part of research conducted between 2017 and 2019 and in a coworking space during the pandemic, we sought to understand who these coworkers are, their uses of these collaboration spaces as well as their assumptions about energy and their visions of the future. The investigation was carried out in three stages. Interviews were conducted with

Fig. 2.2: Sharing values in a coworking space. Photography by Aurore Flipo.

the founders of coworking spaces, then a questionnaire was administered to the users of these spaces, and finally forty interviews were conducted about the coworkers' residential and professional biographies and their uses of coworking spaces.

An ethnography of two coworking spaces was conducted to grasp participants' everyday use. Coworking space C, located downtown in Lyon, is part of a brand that owns four coworking spaces in the city plus several in other cities around France. Its brand sells coworkers an experience of sharing, with everyone who wants to participate, through presentations over lunch and at tea-time. It also organises events ranging from theme-based workshops, meals and sport to drinks. The other coworking space, U, was in the rural context of Drôme. This non-profit organisation uses the vacant space of a factory and aims to contribute to the social, cultural, and economic development of the territory through sharing knowledge and skills. This space welcomes IT workers as well as craftspeople who need workspace. In addition to offering an open space for work, meeting rooms, Wi-Fi, and a printer, kitchen and coffee machine, this coworking organisation holds ecology-focused activities.

Living the coworking experience and digital life

I met Étienne in the courtyard of the C Lyon coworking space, fitted out in a for-mer workshop – in contrast to the others which are located in large bourgeois flats. When I arrived a group of people, who were sitting around tables, sharing a coffee, invited me to join them while I waited for Étienne to arrive. Étienne explained to me that this coworking space is also the only one where ecology is a shared concern, represented in a sign post about the need to turn off the light after leaving particular meeting rooms which do not benefit from natural light, as well as in messages about how to avoid food waste in the kitchen and a collective compost in a corner of the courtyard next to the bicycle parking. This coworking space is also distinguished by its huge hammock that occupies the whole centre space of the second floor below a canopy that sheds natural light onto the open space below where most of the coworkers are working.

Étienne's story and his reasons for working there are very similar to those of most people I met. After studying engineering, he found a job in Paris, working for a large online sales company. Two years later he changed jobs and seized a promotion to live in London, where he spent two years. Tired of the working rhythm, which left him little personal time, he took a new job in Lyon. He had studied in Lyon and decided to move back there because of the more modest size of the city, compared with the megacities, along with its busy nightlife, its climate, the possibility of commuting by bike, and its relative proximity to the Alps. After a year he resigned from his position to create an online platform dedicated to the sale of organic farming products. While moving to Lyon fulfilled his need for a change of pace and direction in his daily life, his new job did not give meaning to his work or fulfil his wish to be part of the creation of a different, less profit-oriented, and less energy-hungry society. This change implied a signif-icant loss of income during the creation of the platform as well as in the long term. Étienne described his professional retraining as ethical, and aligned with his commitments to avoid travelling by car or plane, stop eating meat, and more generally pay attention to the carbon impact of his actions. However, what struck me as a researcher was that he didn't account for the energy costs of his platform, which relies on energy-hungry data centres (see Ch. 4 below).

Working in a coworking space was not an a priori choice but the isolation of working from home did not work for Étienne and he had soon wished to meet people. His new professional partner lived in Annecy, a medium-sized city locat-ed in the Alps where C also offers a coworking space. Choosing C allowed them to have an office and a place to meet outside their homes in both cities as well as an office in Paris when needed, and for Étienne another office in Nantes, where his family lives. Coworking space also seemed a place where they could establish interesting professional relationships. Attracted by the possibilities offered by the digital economy, Étienne is also representative of an ideal of life built through opportunities (Bauman 2008). An individual who requires, to become a subject, to have the capacity to be an actor, to build his existence, to master his experience, and be responsible (Wieviorka 2008). This also implies the ability to adjust his project between contradictory desires and moral injunctions, such as aspirations and material and financial constraints. This coincides with the desires for good lives conveyed by the imagination of lifestyle migrants (Benson and O'Reilly 2009, Cook 2020) offered here by the opportunities of the digital economy, which is perceived as a solution for the future of the planet, rather than as generating its own environmental problems.

Muriel and Olivier's story resonates in some ways with Étienne's, but presents another aspect of the implication of coworking. I met Muriel and Olivier in 2018, and have continued to work with them over two separate projects. They used to live in Paris. Muriel was working for one of Paris's museums, while Olivier worked in industry as an engineer. More and more aware of the impact of his activity on climate change he decided to change jobs in 2008 and work for a non-profit organisation dedicated to developing the energy transition. In 2012, after the birth of their children, they decided to move to the Drôme because of the presence of Montessori schools and the fact that the Drôme Valley is a territory of experimentation of energy and agro-transition. However, Olivier still had a six-hour commute to Paris at least once a week. They chose the city of Crest because of its proximity to one of the high-speed rail stations and the presence of a coworking space; Muriel was then a stay-at-home mother. Two years later, she started to organise events for U, as a volunteer, where she met the creator of an NGO aiming to develop rural sustainable mobility. A few months later she was employed as the coordinator of the NGO. While Olivier is still commuting on a regular basis to Paris and working at U when in the Drôme, Muriel now works for the NGO full time from U, alongside two colleagues.

During the lockdown of spring 2019, U had to close. As a non-profit organisation depending on the office rental income, it was near bankruptcy as coworkers stopped coming and paying. It reopened in July 2019 with a gauge. Over the summer the demand for space stayed low but increased in the autumn and suffered less from the spring 2020 lockdown, since going to work was allowed twice a week. Being out of the way of the major Internet infrastructure, the major difficulty for U has been to meet the increased demand for high-speed connection. This situation was not entirely new but in the aftermath of Covid-19 the situation further deteriorated, mostly due to the increase of on-line meetings. To return to Muriel and Olivier, both already used on-line meetings as much as possible

Figs. 2.3 and 2.4: Rural coworking spaces. Photography by Aurore Flipo.

before the Covid-19 pandemic, but their interlocutors were not necessarily ready to do so. This situation has since changed dramatically and although in-person meetings are possible again, on-line meetings have become a "normal" way to avoid unnecessary car commutes, which is presented as an environmental bonus.

This situation paradoxically gives some "visibility" to the usually invisible infrastructure of the Internet. For many of the coworkers, moving to the Drôme, as for other areas in rural France, involved their assumption that thanks to the Internet it was possible to live everywhere. However, some coworkers already worked in U before the pandemic because its connection was better than the one they could get from their home. When choosing their homes they had focused on factors including proximity to transport axes such as highways or high-speed rail. Ironically, since the pandemic, as they became less dependent on the transport system, coworkers increased their dependency on the digital infrastructure. While this may not discourage people already settled in rural areas, it has discouraged second-home owners who considered moving permanently. It has also created a renewed interest in the small coworking spaces in the centre of the villages where the digital coverage is better than in more remote locations. However, what is still invisible are the energy costs of such delocalisation.

What are coworking energy futures?

Coworking spaces are both emergent from and indicative of the imaginaries and of the future represented by digital capitalism. Presented by its founders and those who have developed it as the future of work, coworking is embedded with contemporary changes in management practices, as well as new expectations about what a good life can be, which in turn stand for a slower and less energy-intensive lifestyle which prefers proximity to long-distance commuting. Coworkers bring new visions of work, they want to choose where they live and with whom and how they work. Moreover, their renewed vision of work settles in a neoliberal framework. Indeed, since coworking spaces are usually chosen on the basis of being where people live or stay, they can also be part of nomadic lifestyles, as offered by C's model, which is designed to meet such needs. Unlike Étienne, many workers did not come from the digital world, rather they all – whether self-employed or employees – shared the desire to find a work environment near their home. The co-workers surveyed were mostly people who had chosen to come to live in Lyon or in the Drôme or had refused to leave the area. Lyon was chosen for these same amenities that attracted Étienne rather than for the vitality of its economic fabric, while those slightly older people moving to the Drôme were looking for a quiet space to raise their family, like Muriel and Olivier.

The Covid-19 pandemic has increased this trend. Homeworking has become the norm for most of the population, for at least part of the week, and the demand for coworking spaces is increasing. Indeed, the search for a work environment perceived as more productive, less distracting, and allowing for a better separation between the private and the professional was and still is one of the main motives of workers who use coworking spaces (Cook 2020, Flipo and Ortar 2020, Orel 2019).

In her ten measures for a sustainable housing presented on 13th October 2021, the French minister of housing stated that along with improved insulation to lower the energy bills, the addition of outside spaces in flats, and bigger windows to allow more light, the creation of coworking spaces should be considered in order to meet new modes and demands of work. When we started the research, coworking was presented as a form of work dedicated to the happy few. It is now a phenomenon of the near present, in which employees will outnumber the self-employed.

Yet these new trends do not change some of the common enduring invisible trends related to coworking. Rather than having similar professions, the common characteristic of the coworkers was and still is their daily dependence on transportation, energy, and digital infrastructures (Internet routes as well as data centres), in order to communicate, store and share data, undertake work activities, or to simply go to work. Indeed, although coworkers have settled to live in chosen locations, these places are not randomly chosen. As the example of the coworking space U suggests, living along secondary roads comes with a cost in terms of access to high-speed Internet. In that respect, cities are still at the nexus of physical and now immaterial flows. However, when work-related energy use is envisaged it is still in relation to transportation and, for people living in rural areas, focused on the car. Work itself is considered as almost neutral in terms of energy, since the work space is shared, the consumption of the laptop is considered as almost insignificant, and the energy costs of storing data are either ignored or presumed to be balanced by the improved efficiency of data centres. The infrastructure and their maintenance needed to move physically and digitally, the nuclear power plants and the data centres required to store data and enable digital work are absent from the imaginary of sustainability surrounding this lifestyle. Technophiles for the most part present technology as something that can and should help find solutions, never as a source of problems. Although some people are anti-digital, those who are pro-digital think that digital technologies can support the ecological transition. More generally, this dimension overlaps with attitudes to technology and the opposition between those who think that technology is the cause of environmental problems and those who think it can provide a solution to those problems, and embodies the renewed relevance of this debate in the context of the current acceleration of digitisation and of the development of 5G, which is going to be even more energy intensive.

Coworking spaces are thus designed to offer workspaces to people who are intended to achieve their aspirations in terms of living environment. Although driven by an ideology of alternative lifestyles, they are also expressions of digital capitalism, and the ultimate avatar of 'nomadism', post-telework. The coworking spaces allow one to leave home and no longer be isolated at home on a daily basis without making long commutes.

Despite the lower energy expenditure linked to daily trips, these ways of working have several energy costs. They need large-scale transport infrastructures for travel to in-person meetings, and require access to Internet infrastructure and data centres in order to provide the digital connectivity needed. The question of energy consumption for professional travel is relatively present in narratives about coworking, where it is presented as a necessary evil linked to the constraints of professional life. In contrast, the question of consumption associated with the use of storage space in the cloud and Internet services is missing. The future is associated with greater freedom of residential choice and the possibility to combine a life on the move with a more settled one (Cook 2020). It is accompanied by a palette of chosen forms of employment, which serve the objectives of digital capitalism while ignoring the forms of consumption that they in turn induce. Nevertheless, cities remain at the core of the possible life choices that characterise this contemporary scenario. Cities are where the nexus of the different types of infrastructure that underpin contemporary work is located. Therefore, two future energy scenarios are possible. In the first cities would become less central, resulting in lowering the energy needs implied by commuting, and making companies more physically dispersed. This would allow for diverse lifestyle choices, but at the price of increased digitalisation, and the energy costs associated with it over a greater area. The second scenario would be a continuation of the impulse arising from the Covid-19 pandemic. This situation would require both digital and transport infrastructures, and demand energy for travel and for digital working from outside the cities.

Energy-demanding expectations: house size, privacy and domestic energy research

Katherine Ellsworth-Krebs

House size and domestic space per person are important determinants for energy demand, largely because increasing space results in more space to heat and/or cool, even as systems become more efficient (Ellsworth-Krebs 2020, Huebner and Shipworth 2017, Lorek and Spangenberg 2019). Yet little is known about if, how, and why people value bigger homes. In this ethnographic case, I offer new insights into the lived reality of house size and domestic space in relation to everyday practices and expectations of what a home is for. In doing so I draw on research which compares how British and Australian households were attracted to their dwellings, their ideal and future homes, and their satisfaction with their current space per person. By exploring the experiences and explanations of living with more space in different cultures, it considers the ways in which space expectations vary based on life stage, temporality, and geography.

I recruited 24 households, half from the UK and half from Australia. I found participants through an agency in order to compare differences based on age (three equal-age cohorts, aged 20 – 30, 40 – 59, and 60 – 80), income (half above and half below national average income) and domestic situations (one-person, two-person, multi-generational household) (Fig. 2.5). Nonetheless, even though I based my sampling strategy on household size and income, the UK households generally lived in half the space of their Australian counterparts. I carried out interviews and virtual home tours in August 2018, involving all members of the household (over the age of 18) together.

I began interviews by asking for descriptions of normal weekday and weekend routines (e. g. where was time spent with others and on their own), what they liked and disliked about their current home. Then participants imagined features of their ideal home, described desired changes to their current home, whether size was an important consideration, and whether they would want a bigger or smaller home in the future. Finally, I asked similar questions about all their previous homes to get a sense of their housing history and how this shaped their current expectations and images of home.

Age (cohort)	Household (income)	House (m²)	Floor area per capita	Household (income)	House (m²)	Floor area per capita	Average: House/Person
		UK			AU		
20–39 (Cohort 1)	1	59	59	1	94	94	UK – 70 /39 AU – 146/71
	2 (low)	72	36	2 (low)	125	63	
	2 (high)	74	37	2 (high)	144	72	
	4	74	25	4	220	55	
40–59 (Cohort 2)	1	30	30	1	55	55	UK – 80/30 AU – 167/81
	2 (low)	100	50	2 (low)	250	125	
	2 (high)	62	31	2 (high)	132	66	
	5	150	30	3	230	77	
60–80 (Cohort 3)	1	45	45	1	70	70	UK – 86/44 AU – 245/114
	2 (low)	70	35	2 (low)	150	75	
	2 (high)	150	75	2 (high)	465	233	
	4	80	20	5	200	40	
Average		79	38		180	79	

Fig. 2.5: Participants by household and house size, age, and country.

'Thirty-four square metre flat and it was, oh, my God, we almost had a divorce'

The vignettes presented here were chosen to offer commonality from a heterogeneous sample with a mixture of household sizes and genders, half each from the United Kingdom and Australia being presented below. Drawing on two Cohort 1 households (20–40 years) highlights the limitations of putting too much stock into planning or policy based solely on what people might imagine they want in the future.

Brazilians Cynthia and Gabriel emigrated to London, UK four years ago. Their second apartment in south-central London was one-bedroom and 60 m², which they described as 'spacious enough' so that they decided to rent out the bedroom and sleep in the living room. After two years, they discovered a flat located centrally which allowed them both to walk to work. Cynthia explained 'we thought it was going to be a good idea. Yeah, let's downsize and not pay the travel card Oyster [£3360/year and actually] go out and enjoy things around.' They imagined an improved quality of living even in a smaller flat: more ease to meet friends, more money to spend eating out, less time commuting. With these imaged benefits they downsized to a one-bedroom 34 m^2 flat near Oxford Street. Yet this turned into a 'stressful home': 'We tried to avoid being at home because we didn't like it and felt very trapped' (Cynthia). Cynthia and Gabriel both enjoy cooking together, but the tiny kitchen led them to 'not want to cook' and instead 'grab something in the street' (Cynthia). Their glamourous and positive image of eating out transformed into something negative in reality. Being unhealthy and eating out became associated with an uncomfortable home. The lack of space in their new flat heightened with visitors staying with them for four out of the twelve months they lived there: 'We are from Brazil, so we have visitors all the time, family and friends. And it was just one bathroom for the whole place and it was chaos' (Cynthia). The number of visitors was higher than previous years, yet with their desire to stay in touch with family and friends from Brazil, having a spare bedroom became an essential part of their vision for an ideal home. As soon as possible, Cynthia and Gabriel moved into their fourth and current, 74 m^2 two-bedroom, two-bathroom new-build flat. Cynthia concluded 'the size now is like perfect for us'. Their experiences are common in the sense that we have to try something to know if it suits us – and they both suggested that under different circumstances or a different layout 34 m^2 could work for them.

While the imagined positives of downsizing did not suit this London couple's needs in the end and led to their moving out of the city centre for more space, an Australian family similarly reflected on downsizing because of their home being 'too big' (Figs. 2.6 and 2.7).

'The other one was just too big'

Abi and Loren are new parents living in a two-storey house with their two- and three-year-old children outside of Melbourne, Australia. Before having children, they completely remodelled a 220 m^2 single-storey detached four-bedroom, two-bathroom house on 750 m^2 of land. They imagined raising children with a huge garden to run around: landscaping, spending all their spare time doing the remodel themselves, calling traders and getting quotes to keep costs down. Yet when children arrived, the house and garden size were less important to their imagined ideal future home than reducing time for house work, maintenance and commuting. The commute became too lengthy – it took 2 hours each way and 'basically we wanted to spend more time with them [our kids]' (Loren,

Figs. 2.6 and 2.7: Abi and Loren's two-storey house outside Melbourne, Australia. Photograph by anonymous research participant, permission granted through consent form.

20 – 40, AU). They downsized in order to afford to move closer to work – now 30 to 40 minutes each way. Their new home has a similar square footage, but feels smaller because it is on two storeys and has one less bedroom. Abi and Loren reflected preferring this smaller home because of less time spent on garden maintenance ('some grass area is OK but you don't want to spend two

hours or more on a weekend just taking care of all that' Abi) and cleaning ('I would like something I can manage in terms of cleaning' Loren). Neither wants as large a home as they had previously in their imagined future, yet their space needs may change as their children get older.

'The size of what I was moving into wasn't relevant at the time, I was moving away from a situation'

Both of the vignettes presented in this section come from Cohort 2 (40 – 60 years old) single households talking about the satisfaction and necessity of living on their own. Similar to the above examples, and public discourse, they emphasised that what attracted them to their homes was a trade-off between the primary factors of location (i.e. reducing commute) or size (i.e. space more expensive towards the city centre). Yet interviews and house tours (as opposed to a short survey) allowed for a more in-depth discussion of motivations, highlighting again the contribution anthropology offers in distinguishing between what people say, do, and say they do. Moving and choosing a home are complex.

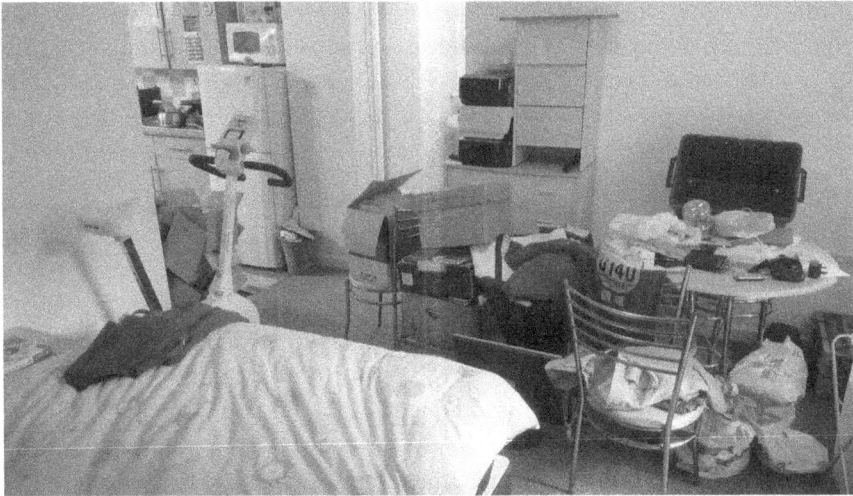

Fig. 2.8: Roger's tiny little flat in Reading, United Kingdom. Photograph by anonymous research participant, permission granted through consent form.

Roger recently separated from his wife, moving out from living with her and his three children: 'so my wife and my kids live there in this lovely big house, and I live in this tiny little flat in Reading' UK. He stressed his choice in a new home as influenced by affordability ('rather cheap' and 'renowned for drug dealers') and central location (e. g. can walk to work and save £1300 on annual commuting costs because 'money's a little bit tighter'). He decided at his age it 'didn't quite appeal' to have to share accommodation, being too much of a 'lottery' living with strangers. Roger's move is still explained in terms of affordability and location, with some implied sadness over it being 'tiny' compared to his previous 'lovely big house.' Yet he also raised the lack of privacy as a contributing factor for needing to move out. Roger reflected on the lovely big home becoming filled and 'cramped', from first moving in with a bean bag and no bed, and sleeping on the floor to expanding the home as the family grew, adding a bedroom and two bathrooms. Nevertheless, he said that no room felt like 'his' and 'when I left, all the rooms were full of children, and stuff and toys.' Nowhere was quiet and that impacted his housing satisfaction: 'one of the things that I didn't like about my previous house, and I struggled with, and now I have what I wanted actually, is just a bit of personal space.' Roger's story is a reminder that home is not fixed: it is a process, in flux, reactive and not always aspirational. The large family home with one's spouse and kids is often seen as 'the ideal', but that does not mean it is always or necessarily comfortable or satisfying.

Barbara moved into her two-bedroom 55 m² flat outside Sydney after living with and caring for her father before he passed away. Initially she mentions being attracted to her current home because of what she could afford on disability payments and inheritance. She also liked the location for its 'sense of life' with cafés, cinemas, and the beach nearby. At the time, though, a key motivation was to have somewhere on her own to recover from her caring responsibilities:

> I basically spent quite a few years recovering from the trauma of looking after him, because it did…things were sort of disappearing in my social life and my sense of self, and what I could do in my own space, and the demands that were placed on me. (Barbara, 40–60, AU)

Barbara moved back to her childhood home to look after her father and even in a home with four bedrooms and a large garden that she had once shared with three sisters and her parents, during this return as a carer she felt she had no space to herself. In the final year, she describes her privacy as one day a week when another carer came or when her father was sleeping. Even then she states she had to be 'constantly with him.'

These four vignettes reveal how everyday practices, such as cooking, cleaning, caring, and commuting, impact energy demand resulting from home heating and mobility. For instance, for Cynthia and Gabrielle (UK), home cooking was a source of pleasure, companionship, and leisure. Their choice to downsize from a 30 m² to 17 m² domestic space per person resulted in the practice of home cooking becoming impossible. The pleasure of the home being a site of cooking disappeared and led them to eat out, spend as little time as possible at home, and created tension between them ('we almost had a divorce'). Socialising or *caring* for family was another key domestic practice shaping their desire for space. With family regularly visiting from Brazil, they would sleep on the sofa in the open-plan living room–kitchen, giving their bedroom to guests, which further undermined the home as a place of rest and privacy. Furthermore their sleep was disrupted by different practices competing for synchronised performance in the same space. Having only one bathroom exacerbated the experience. In other words, moving to a larger home (i.e. 47 m^2 domestic space per person; two-bedroom, two-bathroom flat) was due to the practices and expectations of home that this space enabled: companionship, relaxation, cooking, cleaning, and socialising. In this sense, intervention into reducing space per person requires alternative configurations targeting collective ways of cooking or hosting guests, such as communal canteens and bookable guest rooms in blocks of flats. On the other hand, Abi and Loren were overwhelmed by the time and effort demanded of certain practices (i.e. cleaning and gardening) required in a large home (i.e. 110 m^2 domestic space per person) and once they had children these activities, alongside a two-hour commute, led them to downsize (i.e. 55 m^2 domestic space per person). Thus certain everyday practices, especially in relation to cleaning, tidying, mowing, and commuting, can also impact a desire for (less) space due to the home no longer being experienced as a place of rest.

In relation to energy demand, factors such as commuting distance, transport infrastructure, and house size influencing residential preferences also have clear implications for consumption due to fuel for mobility and heating. Residential mobility describes the process of a household reacting to shifts in their housing needs and preferences and addressing this through moving house (Mulder and Hooimeijer 1999). The vignettes presented here highlight the interconnection between location, especially its impact on commuting, and house size (e.g. less space, shorter commute; more space, more expensive commute). Cynthia and Gabrielle, for instance, went through a process of goldilocks-ing, trialling a walkable commute for a tiny flat in London and finding it unsatisfactory and then moving to somewhere with more domestic space and accepting a longer, more expensive commute. In this way, residential preferences tangle past experiences with imagined futures (e.g. dream of the homely home), yet they depend on the availability of particular housing forms (e.g. detached house). Over the life course, household sizes often decline, as children move out of the family home for instance, and Roger and Barbara are interesting reminders that residential mobility is not always aspirational (e.g. the most central location, the big house with a picket fence). While much academic research and discussion is about the 'ideal' home driving demand for domestic space, and resulting energy consumption, house size and moving also depend on pragmatic considerations such as the dissolution of households ('I was moving away from a situation'). Roger's and Barbara's both choosing to live alone later in life was partly due to a desire for privacy and personal space that they did not get in their previous home due to caring responsibilities for family. There was often an acceptance of cohabiting and sharing accommodation when first leaving the parental house, but later in life it was deemed inappropriate or too much of a 'lottery' to live with strangers. Births, deaths, marriages, divorces, job offers, promotions and employment insecurity were all potential catalysts for moving house and re-evaluating necessary features and affordances of homes.

Participants' explanations of living with more and less space in different cultures reveal the dissonance between their experiences and both their own and broader public assumptions that more domestic space is better. Moreover, the vignettes highlight how domestic space per person, and expectations that the home should offer privacy and personal space (Ellsworth-Krebs, Reid, and Hunter 2020), vary in relation to people's life stage. Trends towards increasing domestic space per person influence and are influenced by our images of future and ideal homes in the sense that the home has a current material existence and yet is hugely shaped by a pursuit of images of the improved future home.

Homes are in flux, a process, a pursuit, and if we continually expect more space, this affects imagined future homes and how homes will continue to evolve. More space per person is a trend that pushes 'normal' life towards being increasingly energy-demanding in high-income countries and should not be overlooked in energy research. Moreover, energy reduction is a systemic issue, not an individual responsibility, and understanding wider societal trends that shape individual's choices and environmental footprint, such as developers creating ever bigger homes or declining household sizes in high-income countries (Ellsworth-Krebs 2020), is essential to designing appropriate interventions.

"I want to be able to open my windows!": Reflections on ownership as a central pillar of energy democracy

Michiel Köhne and Elisabet Dueholm Rasch

Ownership, as one of the pillars of energy democracy, is about the politics of energy: who gets to decide about energy provision and consumption? And how? (Szulecki 2017). In this section we reflect on the different dimensions of ownership over energy decisions in the transition towards a renewable energy future. We argue that different temporal orientations (Bryant 2019) inform, and at the same time become manifest in, different experiences and practices of ownership over energy decisions about consumption and production, which in turn shape different routes towards new energy futures. This argument builds on two central claims regarding lower-income groups' energy practices (Rasch and Köhne 2017): first, their narrow financial margins strongly limit their control over energy decisions, and second, their practices of energy-frugality constitute a way of claiming ownership over energy decisions. We do so by way of a case study of energy practices in the Noordoostpolder (The Netherlands). In what follows, we first briefly discuss the methods that we used during our fieldwork, before we go on to explore how different social groups experience and claim ownership over energy decisions.

Methodology

We collected the material presented in this article between 1 May 2016 and 1 July 2019. The key methods that we used during this period were participant observation, unstructured and semi-structured interviews, and Participatory Action Research. In total we interviewed 25 people that relate in different ways to renewable energy production and consumption. Interviews with renewable energy prosumers, farmers that invested in wind farms, as well as interviews with social housing residents were often conducted prior to, during, or after (or a mix of all these moments) 'walking tours' around houses and businesses related to renewable energy. In addition we conducted participant observation in meetings that discussed ways forward towards a renewable energy future.

The PAR workshops were organised in close collaboration with the social housing corporation Mercatus in December 2018. During three workshops, with 12–15 participants each, we discussed energy practices and experiences

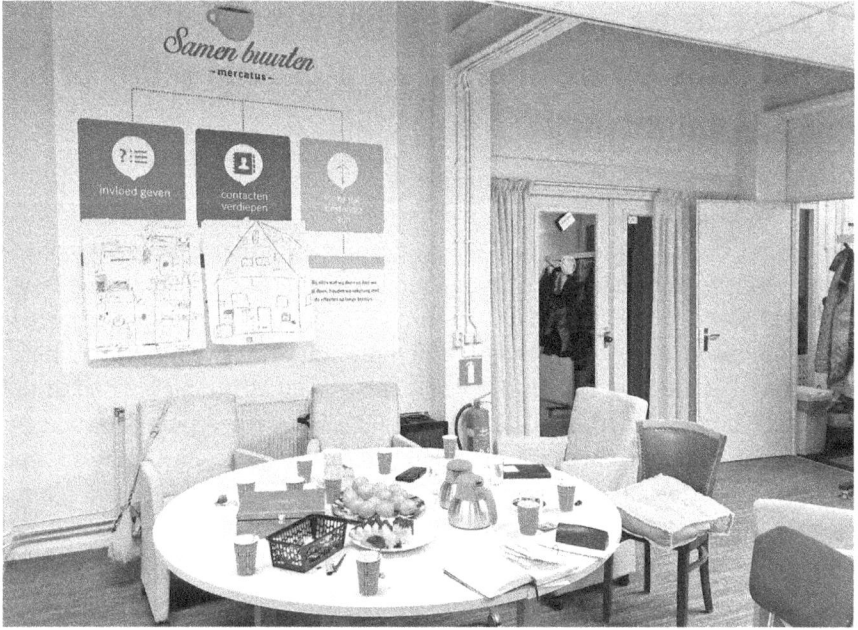

Fig. 2.9: The setting of our PAR workshops with tenants of the social housing corporation Mercatus in Emmeloord (the main town of the Noordoostpolder). Photograph by Michiel Köhne and Elisabet Rasch.

with residents of social housing projects in the Noordoostpolder. We did so by way of a drawing-mapping exercise. We asked the participants to draw a map of their home and indicate their energy practices on these maps. These drawings were the point of departure for our discussions that followed afterwards, where we would invite participants to explain their drawings.

Ownership over carbon-free home-making

Anne has been on social benefits since she was made redundant four years ago, devoting her time to needlework to raise money for the local food bank. Notwithstanding her precarious financial situation, she did save up for an energy-efficient refrigerator. She would also like to have solar panels on her roof, but 'that is such an investment, that is way beyond my means'.

Anne's position illustrates the first tension related to ownership over energy decisions that we identified: the (financial) room for manoeuvre to decide about how to become future-proof in terms of renewable energy. Social housing resi-

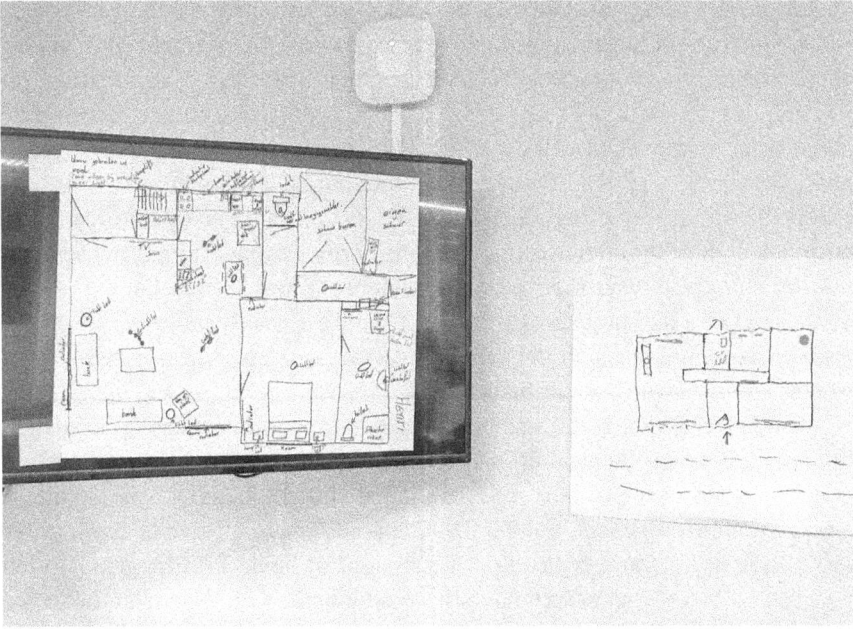

Fig. 2.10: During the PAR workshop, we asked participants to draw a map of their home and indicate their energy practices on these maps. Picture by Michiel Köhne and Elisabet Rasch.

dents can only make small energy investments, and depend on the housing corporation for more radical future-oriented changes, such as extra insulation or solar panels. However, this does not mean that they sit back passively for Mercatus, the social housing corporation in the Noordoostpolder, to step in.

A first way in which social housing residents can claim ownership over their future energy is by informing Mercatus about their needs. Some tenants have had good experiences with this, as is illustrated by Jack: 'I say "well, the wall is freezing cold, something must be wrong", and then they came with a camera and looked into the wall cavity and found that it was completely, hum, collapsed, and then later they fixed it [...].' Mercatus finds it important to align renovations with the needs and wishes of its tenants, but also finds it difficult to involve social housing residents in its energy decisions. Social housing residents often don't find the time to show up at meetings oriented towards the future, or do not respond to letters send out, like Otto: 'Two or three years back we all got a letter [...] in our street nobody replied to that, me neither [...] so, well, then nothing happened, [this] makes sense. Two roads down, they did answer the letter and a lot has been done to their houses.'

Notwithstanding the formal possibilities for participation in energy decisions, many social housing residents express feelings of insecurity and anxiety when it comes to prospective renovations. Often, they do not know when (or if) their houses will be renovated or demolished. In addition, they worry about what a renovation will do to their homes, about the limited ability of a heat pump to heat up their living room, about not being able to open their windows, about draughty living rooms, and about cooking on electric stoves. These anxieties about the unknown future might partly originate from rumours, but they do shape the ways that tenants experience (future) renovations. They are also rooted in a fear of losing control over the ways that tenants can transform their houses into homes. In the end, renovations entail technical improvements whereas tenants worry about their *homes.*

Another way that social housing residents can claim ownership over their future energy use is by making small investments that contribute to smaller energy bills, such as energy-efficient refrigerators and LED lighting. LED lighting is a considerable investment, but worthwhile, according to most tenants we spoke to: 'It uses just 2 watts, with the same amount of light.' LED lamps are often bought one by one, as most tenants have only limited funds available: 'In the end you just do your sums, really, and then you say to yourself, that's all for now.' Mercatus also helps out here. Its energy coaches provide advice on energy use during house visits giving away a few LED lamps as an enticement.

The ways in which social housing residents carve out small niches to claim control over the energy future of their homes contrast sharply with the strategies that several more well-to-do stakeholders in the Noordoostpolder employ to go 'off-grid' and no longer depend on energy providers in the future. This is the case for private homeowners, but also for several large-scale consumers such as greenhouse horticulturists. In addition, Energy Network[1] members consider the local production of renewable energy as an opportunity to gain complete ownership over energy decisions. Especially during the first Energy Network meeting, several members argued that ownership over renewable energy production should be kept away from stakeholders from 'outside the polder', repeating the statement: 'Renewable energy should be of, by, and for the Noordoostpolder' several times. In tune with this line of thought, the Network is setting up membership schemes with relatively low fees for a large solar panel park. However, most social housing residents do not have access to such initiatives because these low fees are still too high for them.

---

1 This Energy Network is the *NETwerk Noordoostpolder Energieneutraal,* the local renewable energy network that was established in 2017 as a platform working towards a carbon–free polder.

Energy-frugality as ownership

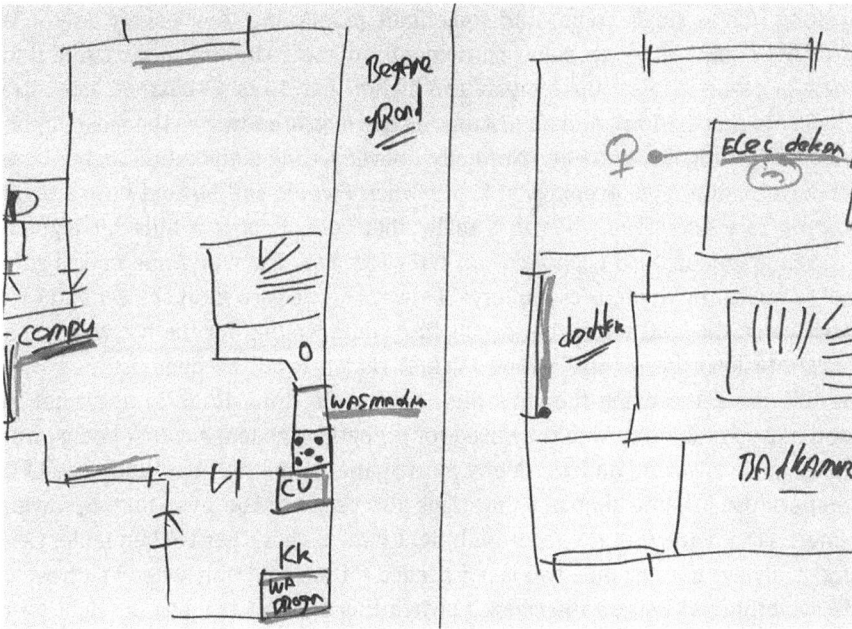

Fig. 2.11: One of the drawings made during the PAR workshop detailing the locations of energy use such as washing machine, computer, and electric blanket. Picture by Michiel Köhne and Elisabet Rasch.

Misha did not choose the house he lives in. It was assigned to him after a very difficult period in his life. He had to start all over again, with almost no resources as he had lost everything: his wife, his job, his house, his car. His economic way of life – recycling water, only using a small gas heater, reusing envelopes to write down groceries – little by little helped him rebuild his life.

Misha's situation illustrates the second tension that we identified and that becomes visible in the two different routes towards a renewable-energy future: everyday-life energy frugality versus producing as much renewable energy as possible with a long-term time horizon. These two routes become visible in the contrasting energy practices and related planning horizons of social housing residents and renewable-energy entrepreneurs.

The most important way in which social housing residents claim ownership over energy decisions is through everyday energy frugality. Saving energy means saving money. Tenants employ a plethora of frugality tactics informed by a short-term temporal orientation and limited resources. The most important one is saving on heating. Almost all tenants that we spoke to looked for ways to keep the thermostat as low as possible while still feeling comfortable; using blankets, heating only a single room, and sometimes monitoring their energy use on a monthly basis. Although some participants in the PAR workshops agree that there is no need to set the temperature higher than 16 or 17 degrees, most use slightly higher settings, and all discuss the thermostat setting as the most important energy-frugality strategy: 'Normally I never set the temperature higher than 18 degrees and in the evenings at 19.5 ... when I would still be cold I use a plaid and on rare occasions when it is really that cold, I set it a little bit higher.'

Many tenants also economise on hot-water use. The way Anne makes sure not to waste any water, is exemplary: "I always take a very short shower and I always catch the cold water that comes first while waiting for the hot water in a bucket to feed the plants." When Michiel visited Peter, he demonstrated how he fills the kettle using the cups that they would drink from, in order not to heat a drop more than would be used for preparing the tea. Another energy-frugality tactic is saving on light. Many participants of the PAR workshop use LED lamps or use just one lamp at a time. This also became clear in an interview with Ginger, who pointed at the three-bulb light fixture above her kitchen table, saying: "I always use just one, two is not needed for me, and only when I sit here to do something, if I sit on the couch I switch this light off and use the light over there." Saving energy is part of many different aspects of tenants' everyday life and as such constitutes important ownership over energy decisions.

These energy frugality tactics emerge from a short-term temporal orientation in which the everyday reality of running a low-income household, going from day to day and from month to month, limits planning horizons. For some tenants energy frugality is more than a way to make ends meet; they find it important to contribute to climate-change mitigation, and to look further ahead, like Anne: 'You hope that your grandchildren will also be able to have a nice life and this may cost a bit extra.' However, most tenants discuss energy-frugality tactics primarily as a way to make ends meet.

In contrast to claiming ownership over energy decisions through energy frugality, stakeholders in the Energy Network (re)appropriate ownership over energy decisions by way of producing as much renewable energy as possible. Such energy production in the Noordoostpolder is primarily driven by farmers who seek livelihood security through diversification and is characterized by large investments and long-term planning. In tune with this point of view, most energy decisions are geared towards producing enough energy to maintain a comfortable lifestyle, in contrast to saving energy and adapting to a less luxurious lifestyle. For them, monitoring energy use is less a worry about monthly payments and more about the pleasure of witnessing how their investments transform into profit.

Fig. 2.12: Solar panels on top of the town hall. Picture by Michiel Köhne and Elisabet Rasch.

Discussion

In this case we examined how social housing residents experience and claim ownership over energy decisions and how this is informed by temporal orientations, contrasting this with well-to-do home owners and renewable energy entrepreneurs. Energy transition policies are often rooted in long-term temporal orientations. In line with such policies, proposed renovations for a future-proof housing stock often anticipate similarly distant futures. In addition, policies tend to focus on technological fixes, rather than on processes and practices of home-making that are important for social housing residents. Such energy policies are in tune with the temporal orientations of renewable energy entrepreneurs and more well-to-do homeowners. Social housing residents, however, often experience anxiety and insecurity about energy-related renovations. Living from day to day, their main way of claiming control over energy decisions is by practising 'energy frugality'.

Temporal orientations towards the future inform (everyday) energy practices and experiences of ownership in energy decisions. In Bryant's words: 'Whether fleeting moments or the result of long-term planning, whether individual feelings or part of a collective vernacular, we are constantly anticipating, expecting, hoping for, and speculating about – and thus living – the future' (Bryant 2019: 4). In the case of the Noordoostpolder, socio-economic positions shape the ways that people 'live the future'. Tenants' short-term temporal orientation limits their ownership over energy to decisions about small home-energy improvements and everyday energy frugality. More well-to-do people install state-of-the-art renewable energy equipment and need only to wait to see their energy costs diminish over a longer period of time. Renewable-energy entrepreneurs work with time horizons up to ten years and more, claiming full ownership over local energy production. As a consequence, social housing residents often feel excluded from (decision-making processes related to) the energy transition towards renewable energy.

Fig. 2.13: Large-scale energy production by the Noordoostpolder Windpark. Picture taken by Michiel Köhne and Elisabet Rasch.

Social housing residents' perspectives on ownership over energy decisions offer some important insights about the meaning of "ownership" for energy democracy. Energy democracy as an ideal builds on the idea that renewable energy's potential for decentralised production could foster decentralised ways of governing energy (Burke and Stephens 2017, 2018). In this line, scholars that work on energy democracy question who controls energy, to what end, and to whose benefit (Fairchild and Weinrub 2017).

Our research with tenants in the Noordoostpolder shows that marginalised groups with little resources to spend and, consequently, short-term temporal orientations often feel excluded from energy decisions that affect their homes. They voice the desire to be able to decide on small home improvements, such as being able to open windows, to be better informed about what will happen, and above all to know for sure that it will not cost them more. For social housing residents, dealing with the future of energy means economising in the present. Limited resources, as well as short-term temporal orientations (which, in turn are rooted in these limited resources), limit the options for tenants to invest in, and thus profit from, renewable-energy technologies. In addition, the social housing corporation decides in the end what will happen to their homes. Hence social housing residents do not own renewable energy technologies and their participation in energy decisions is often limited.

Our research offers two possible entry points for including social housing residents more explicitly in participatory governance of energy: ownership over home improvements and ownership over energy use. Both dimensions are important ways for tenants to turn their houses into homes and at times contrast with housing corporations' (ideas for) investments and renovations that seem to be more about house improvement and technological fixes than about home improvement. Information that fits tenants' daily living circumstances as well as spaces for participation that suit their daily routines could both contribute to more inclusive ways of governing energy. Tenants take control over energy decisions through day-to-day frugality. These day-to-day savings are a near-future-oriented energy practice and, although prompted by limited resources, do contribute to a fossil-free future. Hence, energy frugality, a way of dealing with day-to-day energy challenges, rooted in a short-term temporal orientation, brings the distant, undefined fossil-free future into the present. Being able to control the use and the costs of energy is an important dimension of ownership for social housing residents.

These dimensions of ownership are very different from the ways that renewable energy entrepreneurs and the powerful Energy Network claim ownership over energy decisions. Their talk of long-term planning, energy production, and comfortable lifestyles excludes groups with limited financial resources and short-term planning horizons from their vision and their policies. As such, the transition towards a renewable energy future reproduces categories of exclusion: social housing residents do not benefit from the advantages of renewable energy, like comfort and lower energy bills, or even an extra income.

Temporal orientation and ownership

In sum, our case shows how limited (financial) resources and short-term temporal orientation towards the future shape experiences of ownership over energy decisions. In general, housing corporations decide about technical fixes to make their housing stock future proof. This can cause feelings of anxiety and insecurity among social housing residents. In addition, tenants have limited resources to invest in renewable energy technology. In contrast, more privileged groups and renewable energy entrepreneurs have long-term temporal orientations and claim complete control in energy decisions to go off-grid. The most important way for social housing residents to claim control over energy decisions is through energy frugality. We conclude that the democratising potential of renewable energy production technologies for social housing residents lies in taking into account these two dimensions of ownership when developing policies regarding future-proof social housing.

Next steps

The cases presented above surface multiple ways that the everyday complicates dominant narratives about energy futures. In turn they indicate new ways forward through attention to the alternative narratives that emerge from everyday experience and imagination. We draw attention to three key points:

The everyday complicates dominant future visions in each of the cases. We see how people's priorities, routines, and practices, and abilities to improvise, all participate in directing energy futures along particular routes. The cases revealed this in several ways. We learned how dominant narratives are contested in the present as people live out everyday practices of anticipation. For example, Köhne and Rasch contrast orientations to the future in the everyday life of social housing tenants with those of more privileged groups and renewable energy entrepreneurs, as an example of how techno-solutionist future visions diverge from everyday life energy futures. This complicates dominant policy narratives by critiquing the ways in which such tenants' own energy futures remain unseen, ignored, and not built upon by policy makers, thus exemplifying a disregard for energy democracy. Ortar and Flipo unravel the entanglements of working location and lifestyles choices. The implications on the future of energy consumption are multiple: if less energy is spent on everyday transport, potentially negatively impacting the availability of public transport and exacerbating inequalities for those who can't make such choices, home sizes will increase due to the need for home-offices; allowing more liberty in the choice of where to live will increase the need for everyday car travel to reach working and coworking spaces in response to increased demand for digitalisation in combination with a need for longer distance travel to reach head-office. Dahlgren and Pink explore the likelihood of future industry visions for EV futures being played out in everyday futures, where people have particular priorities and ways of being and living that they are not prepared to give up. As their case shows, however much 'sense' future scenarios might make to people, this doesn't mean that they consider them viable for their own future lives.

Everyday uncertainties and anxieties also form part of the anticipatory modes of life through which energy futures are shaped. Uncertainty is an essential element of our human condition, and no less so in everyday life. We found that people across the diverse sites of our research coped with uncertainty in different ways. For instance, ways of living with and controlling uncertainty can include everyday energy frugality, charging an EV at a particular time of the day because you don't know when you will need it, or coping with the unreliability of the Internet when working from home and having increased dependency on digital infrastructures but using a coworking centre. One way to research and explore uncertainties is to develop experimental methods, which take people out of the temporalities in which they feel they know what is likely to happen next, to open them up to new possibilities (Akama, Pink, and Sumartojo 2018). As Dahlgren and Pink's case showed, when confronted with uncertain futures people often still hold on to their priorities, and to the values that underpin their everyday actions.

People plan for their everyday futures in particular ways which shape the possibilities for energy use that open up for them. Life trajectories usually shape up as personal or household projects, as people consider where they would like to live, in what kind of home, and with what amenities nearby. The modes of life planning that this involves also means that people look ahead to their everyday futures. As Ellsworth-Krebs's case showed, the expectations of home and moving house has energy implications, especially from space heating and cooling, which is influenced by everyday considerations that were highlighted in occupant's previous homes. Indeed, sometimes the acquisition of the idealised big family home (Dowling and Power 2012) is revealed not to be as satisfying in reality. Too much gardening, a long commute, or a breakdown in familial relationships can lead to a desire for smaller homes or households later in life which might be unexpected when home is so often imagined and researched in predominantly abstract and aspirational ways (Brickell 2012). The shifts and changes in life also participate in shaping energy demand, as people might begin to work from home, create new transportation needs shapes, and generally re-shape the way life is configured.

Ethnographic research draws our attention to the everyday inequalities that are so fundamental to the concerns of social scientists. The solutions proposed in dominant narratives tended to focus on the privileged lives of middle-class households, and men who commute to work. The cases questioned the ability of technological solutions to suppress inequalities, and invite us to consider questions of whose everyday life even affords access to EVs, spacious housing, moving out of the city to start a new life in the countryside, having the luxury to choose to work from home or from a coworking centre or producing or accessing renewable energy.

Conclusion

As we write, responses to the multiple issues that we face when we think about futures – including public health, climate and energy transition – are emerging, including in the sites of everyday life. These responses are still shaping aspects of energy futures. The ways of living, working, and playing that are still emerging are also crucial to understanding how energy futures can possibly be imagined, in everyday or institutional contexts. Yet, as existing studies show clearly, industry visions of futures are often misaligned with the possible futures that are revealed by ethnographic research. This makes it all the more important to follow our focus in this chapter on everyday life in homes, mobilities, and coworking spaces as it evolves through and 'after' the pandemic and engages with the climate crisis; in this chapter we offer a starting point, to make it clear what the direction of research and action, as we move into uncertain futures needs to be. Indeed in an ongoingly emerging world, we cannot speak of endpoints, but rather of approaches with which to move forward.

Acknowledgements

We thank all the people who have participated in our research, since without their collaboration and commitment it would have been impossible to develop the cases discussed in this chapter. The research we discuss in this chapter was supported by the following funding organisations and research partnerships: Kari Dahlgren's and Sarah Pink's research was supported by the Australian Government through the Australian Research Council's Linkage Projects funding Scheme ('Digital Energy Futures' project number LP180100203) in partnership with Monash University, Ausgrid, AusNet Services, and Energy Consumers Australia; Nathalie Ortar and Aurore Flipo's research was funded by the French National Research Agency (ANR) (project number: Projet-ANR-17-CE22–0004); Katherine Ellsworth-Krebs's research was funded by the Carnegie Trust Research Incentive Grant (Grant no. RIG007515) and a visiting fellowship award from RMIT University's Beyond Behaviour Change Research Group.

References

Akama, Y., Pink, S., and Sumartojo, S. (2018). *Uncertainty and possibility*. London: Bloomsbury.

Amit, V. (2000). *Constructing the field: ethnographic fieldwork in the contemporary world*. London: Routledge.

Bauman, Z. (2008). *The art of life*. Cambridge: Polity Press.

Benedetto-Meyer, M., and Klein, N. (2017). Du partage de connaissances au travail collaboratif: Portées et limites des outils numériques. *Sociologies pratiques*, 34(1), 29–38. Cairn.info. https://doi.org/10.3917/sopr.034.0029.

Benson, M. C., and O'Reilly, K. (2009). *Lifestyle migration: expectations, aspirations and experiences*. Farnham: Ashgate Publishing, Ltd.

Bille, M. (2018). *Homely atmospheres and lighting technologies in Denmark: living with light*. London: Bloomsbury.

Brickell, K. (2012). 'Mapping'and 'doing' critical geographies of home. *Progress in human geography*, 36(2), 225–244.

Bryant, R. and Knight, D. M. (2019). Orientations to the future: an introduction. In Rebecca Bryant and Daniel M. Knight (eds.), *Orientations to the future*. https://american ethnologist.org/features/collections/orientations-to-the-future

Burke, M. J., and Stephens, J. C. (2017). Energy democracy: goals and policy instruments for sociotechnical transitions. *Energy research and social science*, 33, 35–48.

Burke, M. J., and Stephens, J. C. (2018). Political power and renewable energy futures: A critical review. *Energy research and social science*, 3578–3593. https://doi.orghttps://doi.org/10.1016/j.erss.2017.10.018.

Cook, D. (2020). The freedom trap: digital nomads and the use of disciplining practices to manage work/leisure boundaries. *Information technology and tourism*, 22, 355–390.

Couldry, N., and Mejias, U. A. (2019). Data colonialism: rethinking Big Data's relation to the contemporary subject. *Television and new media*, 20(4), 336–349. https://doi.org/10.1177/1527476418796632.

CSIRO, nd. About-CSIRO. https://www.csiro.au/en/about.

Dahlgren, K., Pink, S., Strengers, Y., and Nicholls, L. (2022). Co-presence and contingency: comics as a methodological innovation in researching automated futures. *Qualitative inquiry*. https://doi.org/10.1177/10778004221097630.

Dahlgren, K., Strengers, Y., Pink, S., Nicholls, N., and Sadowski, J. (2020). Digital Energy Futures: review of industry trends, visions and scenarios for the home. Emerging Technologies Research Lab (Monash University). Melbourne, Australia. https://www.mon ash.edu/__data/assets/pdf_file/0008/2242754/Digital-Energy-Futures-Report.pdf.

Dahlgren, K., Pink, S., Strengers, Y., Nicholls, L., Sadowski, J. (2021). Personalization and the smart home: questioning techno-hedonist imaginaries. *Convergence*, August 2021. doi:10.1177/13548565211036801.

Dányi, E., Spencer, M., Maguire, J., Knox, H., and Ballestero, A. (2021). Propositional politics. In J. Maguire, L. Watts, and B. R. Winthereik (eds.), *Energy worlds in experiment*. Manchester: Mattering Press. 66–94.

Deloitte (2018). Energy accelerated: a future focused Australia. https://content.deloitte.com. au/20180824-ene-inbound-energy-accelerated-registration?_ga=2.201387857. 1767206790.1632981964-533022645.1632981964.

Dowling, R., and Power, E. (2012). Sizing home, doing family in Sydney, Australia. *Housing studies*, 27(5), 605–619.

Ellsworth-Krebs, K. (2020). Implications of declining household sizes and expectations of home comfort for domestic energy demand. *Nature energy*, 5, 20–25.

Ellsworth-Krebs, K., Reid, L., and Hunter, C. J. (2019). Integrated framework of home comfort: relaxation, companionship and control. *Building research and information*, 47(2), 202–218.

Eubanks, V. (2018). *Automating inequality: how high-tech tools profile, police, and punish the poor.* London: Picador, St. Martin's Press.

Fairchild, D., and Weinrub, A. (2017). Energy democracy. In *The community resilience reader*, 195–206. New York: Springer.

Flipo, A., and Ortar, N. (2020). Séparer les espaces pour maîtriser le temps. La reconstruction des barrières temporelles et spatiales entre vie privée et vie professionnelle par le coworking. *Temporalités. Revue de sciences sociales et humaines*, 31–32. http://journals.openedition.org/temporalites/7712.

Graham, P., and Havas, L. (2019). Projections for small scale embedded technologies. CSIRO. Australia. https://aemo.com.au/-/media/Files/Electricity/NEM/Planning_and_Forecasting/ Inputs-Assumptions-Methodologies/2020/CSIRO-DER-Forecast-Report.

Hargreaves, T., Wilson, C., and Hauxwell-Baldwin, R. (2018). Learning to live in a smart home, *Building research and information*, 46(1), 127–139. doi:10.1080/09613218.2017.1286882.

Henning, A. (2005). Climate change and energy use: the role for anthropological research. *Anthropology today*, 21, 8–12. https://doi.org/10.1111/j.0268-540X.2005.00352.x.

Huebner, G. M., and Shipworth, D. (2017). All about size?–The potential of downsizing in reducing energy demand. *Applied energy*, 186, 226–233.

Johnson, C. (2020). Is demand side response a woman's work? Domestic labour and electricity shifting in low income homes in the United Kingdom. *Energy research and social science*, 68, 101558. https://doi.org/10.1016/j.erss.2020.101558.

Knox, H. (2021). Climate change and the politicisation of the mundane. In J. Maguire, L. Watts, and B. R. Winthereik (eds.), *Energy Worlds in Experiment*. Manchester: Mattering Press.

KPMG (2018). Electric vehicles: is the energy sector ready? Available at https://assets.kpmg/
content/dam/kpmg/au/pdf/2018/electric-vehicles-is-the-energy-sector-ready.pdf.

Leducq, D. (2021). Après un an de crise, quelles perspectives pour les espaces de coworking?
The Conversation. https://theconversation.com/apres-un-an-de-crise-quelles-per
spectives-pour-les-espaces-de-coworking-157069.

Lorek, S., and Spangenberg, J. H. (2019). Energy sufficiency through social innovation in
housing. *Energy policy*, 126, 287–294.

Markham A. (2021). The limits of the imaginary: challenges to intervening in future
speculations of memory, data, and algorithms. *New media and society*, 23(2), 382–405.
doi:10.1177/1461444820929322.

Michael, M. (2016). Speculative design and digital materialities: idiocy, threat and
com-promise. In S. Pink, E. Ardevol, and D. Lanzeni (eds.), *Digital materialities:
anthropology and design*. London: Bloomsbury.

Monnoyer-Smith, L. (2017). Transition numérique et transition écologique. *Annales des mines
– responsabilité et environnement*, 87(3), 5–7. Cairn.info. https://doi.org/10.3917/re1.
087.0005.

Morozov, E. (2013). *To save everything click here: technology, solutionism and the urge to fix
problems that don't exist*. London: Penguin.

Mulder, C. H., and Hooimeijer, P. (1999). Residential relocations in the life course. In
Population issues. Dordrecht: Springer. 159–186.

Orel, M. (2019). Coworking environments and digital nomadism: balancing work and leisure
whilst on the move. *World Leisure Journal*, 61(3), 215–227. https://doi.org/10.1080/
16078055.2019.1639275.

Ortar, N. (2018). Mobile/immobile: Qu'apporte le télétravail aux familles de grands mobiles?
In C. Imbert, E. Lelièvre, and D. Lessault (eds.), *La famille à distance*. Aubervilliers:
Éditions de l'INED. 293–308.

Ortar, N., and Ryghaug, M. (2019). Should all cars be electric by 2025? The electric car
debate in Europe. *Sustainability*, 11(7). 1868.

Pink, S. (2022). Trust in automation. In S. Pink, M. Berg, D. Lupton, and M. Ruckenstein
(eds.), *Everyday automation: experiencing and anticipating emerging technologies*.
London: Routledge.

Pink, S., and Leder Mackley, K. (2015). Social science, design and everyday life: refiguring
showering through anthropological ethnography, *Journal of design research*, 13(3),
278–292.

Pink, S., and Postill, J. (2019). Imagining Mundane Futures. *Anthropology in action*, 26(2),
31–41. https://doi.org/10.3167/aia.2019.260204.

Pink, S., Fors, V., and Glöss, M. (2018). The contingent futures of the mobile present:
automation as possibility, *Mobilities*, 13(5), 615–631.
doi:10.1080/17450101.2018.1436672.

Pink, S., Fors, V., and Glöss, M. (2019). Automated futures and the mobile present: in-car
video ethnographies. *Ethnography*, 20(1), 88–107.

Pink S., Ferguson, H., and Kelly, L. (2022). Digital social work: conceptualising a hybrid
anticipatory practice. *Qualitative social work*, 21(2), 413–430.
doi:10.1177/14733250211003647.

Pink, S., Ruckenstein, M., Berg, M., and Lupton, D. (2022). Everyday automation: setting a research agenda. In S. Pink, M. Berg, D. Lupton, and M. Ruckenstein (eds.), *Everyday automation: experiencing and anticipating emerging technologies*. London: Routledge.

Rasch, E. D., and Köhne, M. (2017). Practices and imaginations of energy justice in transition. A case study of the Noordoostpolder, the Netherlands. *Energy policy*, 107, 607–614.

Sadowski, J., and Levenda, A. M. (2020). The anti-politics of smart energy regimes. *Political geography*, 81, 102202.

Sajous, P. (2019). Le télétravail: Sur la voie de la banalisation? Étude à partir d'un cas de télétravail à temps complet (Soho solo, Gers) et d'un cas de télétravail à temps partiel (Safran Nacelles). *Espace population sociétés*, 2. https://doi.org/10.4000/eps.9089.

Schelhas, J., Hitchner, S., and Brosius, J. P. (2018). Envisioning and implementing wood-based bioenergy systems in the southern United States: Imaginaries in everyday talk. *Energy Research & Social Science*, 35, 182–192. https://doi.org/10.1016/j.erss.2017.10.042.

Schreier, M. (2012). *Qualitative content analysis in practice*. London: SAGE.

Smith, R. C., and Otto, T. (2016). Cultures of the Future: Emergence and Intervention in Design Anthropology. In R. C. Smith, K. T. Vangkilde, M. G. Kjærsgaard, T. Otto, J. Halse, and T. Binder (eds.), *Design anthropological futures*. London: Bloomsbury Academic. 19–36.

Strengers, Y. (2013). *Smart energy technologies in everyday life*. Basingstoke: Palgrave MacMillan.

Strengers, Y., and Kennedy, J. (2020). *The smart wife: why Alexa, Google Home and other smart home devices need a feminist reboot*. 1st edn. Cambridge, MA: MIT Press.

Strengers, Y., Pink, S., and Nicholls, L. (2019). Smart energy futures and social practice imaginaries: forecasting scenarios for pet care in Australian homes. *Energy research and social science*, 48, 108–115.

Strengers, Y., Dahlgren, K., Nicholls, L., Pink, S., and Martin, R. (2021). Digital energy futures: future home life. Emerging Technologies Research Lab (Monash University). Melbourne, Australia. https://www.monash.edu/__data/assets/pdf_file/0011/2617157/DEF-Future-Home-Life-Full-Report.pdf.

Strengers, Y., Dahlgren, K., Pink, S., Sadowski J., and Nicholls, L. (2022). Digital technology and energy imaginaries of future home life: comic-strip scenarios as a method to disrupt energy industry futures. *Energy research and social science*, 84, 102366.

Szulecki, K. (2018). Conceptualizing energy democracy. *Environmental politics*, 27(1), 21–41.

Velkova, J., Magnusson, D., and Rohracher, H. (2022). 'Smart thermostats and the algorithmic control of thermal comfort', in S. Pink, M. Berg, D. Lupton, and M. Ruckenstein (eds.), *Everyday automation: experiencing and anticipating emerging technologies*, 171–184. Abingdon: Routledge.

Wieviorka, M. (2008). *Neuf leçons de sociologie*. Paris: Éditions Robert Laffont.

Wilhite, H. (2005). Why energy needs anthropology. *Anthropology today*, 21(1–2). doi:10.1111/j.0268-540X.2005.00350.x.

Simone Abram, Chiara Bresciani, Hsin-yi Lu, Katja Müller, and Asta Vonderau

3 Contested futures of/with energy generation

Introduction

In this chapter we look at how the futures of and with energy technologies are imagined by those who work with them, and by those who engage with them in turn. How are futures contested in relation to different technological trajectories, and who is involved in shaping dominant narratives and forms of resistance? This chapter aims to include serious attention to the materialities of energy infrastructures as well as the politics and socialities of energy supply.

There is now a substantial body of literature in anthropology about life in the energy industries. In relation to oil extraction, in particular, there are significant ethnographies on oil worlds and their impact on the lives of people working in the industry. Shever (2012) traces the lives of oil workers in Argentina, showing how the nationalisation and privatisation of oil industries relied heavily on the labour of workers beyond corporations. She shows how the relations between worker and employer were framed in the language of kinship, notably the paternal role of the company in securing the well-being of employees and their families, as well as the significance of kinship in accessing employment among the families of workers. Breglia (2013) and Canino and Strønen (2015), while also focusing on workers, show fishing communities on the Yucatán coast of Mexico 'waiting for the promises of plenty' amid the oil boom of the late twentieth and early twenty-first centuries (Breglia 2013: 3).

But the significance of anticipation associated with future imaginaries of energy worlds is most clearly articulated in Weszkalnys's work on the elusive oil boom in the West African islands of São Tomé and Príncipe. Weszkalnys shows how expectations alone brought real, material and political changes to São Tomé and Príncipe, without any actual oil being extracted (2011, 2014, 2016). Oil promised to transform a derelict plantation economy into what Weszkalnys calls 'an economy of expectations' (2011: 349) – a concept that links all of the case studies in this chapter – based on oil's function as a saviour, one which she argues held 'an astounding grip on popular and scholarly imaginations' (2011: 350). This imaginative future linked resources to extreme affect: the 'euphoria, excitement, aggression, doubt, trepidation, frustration, disillusionment, and so on [that] regularly emerge, successively or alongside each other, in con-

https://doi.org/10.1515/9783110745641-004

texts of resource prospecting and extraction.' (2016: 128). Along with that affect and the expectation of resources flowed enormous currents of capital, from both private and state institutions. Yet this flow of money was also seen by commentators as dangerous, a potential threat to the development of the islands, and, as such, a risky path towards an otherwise desired future.

Uncertainty about what the future might bring, and disagreements about what kind of future is desirable are often the basis for friction around impending developments (Abram and Weszkalnys 2013). For example, Michael Cepek's lively book, *Life in oil* (2018), shows how Cofán people in Ecuadorian Amazonia navigate the loss of territory and pollution from oil spills that have engulfed their territory over the last half century. In contrast to more photogenic groups in the Amazon forest, the Cofán whom Cepek writes about are often described as already corrupted, with some having worked for the oil industry themselves, and many eking out a livelihood in the cash economy. Some authors even go as far as to claim they are symbols of the deculturation and pollution – symbolic as well as environmental – and even extinction that threaten more pristine communities (Cepek 2018, 50). But Cepek writes with sensitivity and without prejudice of the ways that people can continue to question and reimagine the futures offered by the oil industries, to create meaningful lives, and to challenge Western fascination with cultural purity and indigeneity.

The Cofán of Cepek's description share a predicament with very many people around the world, in wanting to define their own future, rather than live with one imposed by extractive corporate interests. Sometimes these chime with the ambitions of global environmental activists and their frequent desire to promote indigenous or nature-close livelihoods, and sometimes they do not (e. g. Abram 2016). In some cases it does not take an environmentalist to identify the explicit racism of oil industry enclaves. Appel has described in detail the 'infrastructural violence' entailed in the organisation of hydrocarbon extraction in Equatorial Guinea (2012). Private luxury in the enclaves of expat workers, with their suburban homes amid manicured lawns, their reliable and generous electricity and US telephone numbers, contrasts shockingly with the barracks to which South Asian workers are confined, the shared dormitories for Filipino workers, and the prohibition on Equatoguineans entering the compounds at all. Appel describes the exaggerated aspects of private luxury as 'securitised monuments' to escalating petro-inequality and the failure of the state to invest in public services (2012: 442). The corporations' sponsorship of everyday state violence through worker control, suppression of labour organisations, and racially based zoning, is accomplished by maintaining a repressive presidential regime, enabling the companies to justify the security that affords their workers excessive luxury and privilege. The enclaves keep alive a colonial-era plantation mentality, with workers able to remain 'disentangled' from the state and its people, while the oil industry cannot but be deeply entangled in the social, legal, political, and environmental context in which it operates. In investing so heavily in maintaining their own infrastructural regime, the corporations prevent the population from redefining their future according to their own priorities. Owning the future is a way to own the present and crush dissenting views.

This most explicit form of extractive practice is well documented in relation to oil and other mining activities (e.g. Gilberthorpe and Rajak 2017). Anthropology of extractive industries has developed since June Nash's famous *We eat the mines and the mines eat us* from 1979 in what Rajak and Gilberthorpe describe as an important shift. Recent work, such as that inspired by Nader's work in California (1972), focuses on industries and governments themselves, tracking 'the transnational flows and corporate practices of resources extraction' (Gilberthorpe and Rajak 2017: 190). Several authors have picked up Foucault's various analyses of power – whether biopower or governmentality – to note that energy infrastructures are similarly embedded or saturated in structures of political power and normative moralities (Boyer 2014, Dale 2016, Loloum, Abram, and Ortar 2021).

In general terms, the ability to control energy economies relies on the ability to dominate the narrative about what energy is, and what it means for the future, whether that is framed in terms of hopes for prosperity and comfort, or fears of unemployment and deprivation. As Stewart explains (2016), one way to achieve domination of the narrative is by co-opting familiar forms, such as visual media, or by popularising media that promote a world-view and future-orientation that suits the political agenda. It is not only the promotion of such materials that creates effect, but a way of seeing the world that it supports, which Stewart calls 'extractive seeing', a stance that affects renewable as much as it does fossil-fuel energy practices. It reinforces political arguments by normalising one way of seeing the world and lending it persuasive authority that cuts through counter-narratives and alternate, particularly subaltern voices, as has been repeatedly pointed out in critical development studies (Ferguson 1994, Escobar 1995).

It is relatively easy to understand resource capitalism as a form of extractivism, which Acosta suggests is the central mechanism of 'colonial and neo-colonial plunder and appropriation' (2013: 63). Howe has argued that extractivism, as a colonial mentality, applies to immaterial resources such as the power of the wind (2019). For wind turbines to work effectively, or 'optimally' (in a technical sense), it makes sense to locate them in the windiest locations which are more likely to be places considered from the centres of power to be 'remote': the low-lying isthmus of Tehuantepec, or the mountainous rural areas of Catalonia (Franquesa 2018), or across rural Greece (Argenti and Knight 2015). Franquesa outlines how Catalonia has long been used as a resource base for Spain, leaving rural inhabitants facing economic hardship as their agricultural activities are progressively devalued. Just as farmers in the region began to see success by re-framing their enterprises as organic destinations, bringing tourists to enjoy the bucolic landscape, the government proposed to build extensive wind farms over the most picturesque valleys. Seen from a historical perspective, the repeat-ed experience of resource extraction – water, nuclear power, hydro-electric power – and the history of independence struggles, Catalan resistance to wind power is understandable. Far from their being Luddites, set against change or climate change deniers, Franquesa puts forward a Marxist argument that Catalan resistance falls into a conflict between the ability of people to reproduce their livelihoods and define their own energy futures, and the drive of capitalists to produce 'for accumulation' (Franquesa 2018: 20).

With the pressure to decarbonise energy supplies in the global North, renew-able resources in the global South are being eyed up for a new form of extrac-tion. European countries, and the EU, have repeatedly imagined the Sahara as a source of solar power for export to Europe, for example, just as US corporations have established wind farms in Mexico to power manufacturing for export, and hydro-electric mega-dams have been used as tools of geopolitics (e.g. Jensen 2019, Stevenson and Kamski 2021, Gore 2017). In other words, despite the encour-aging rhetoric around renewable energy being a valuable opportunity for dis-persed energy generation and local provision, renewable energy is as susceptible to extractive mentalities as other, more material resources. How technologies are imagined to contribute towards diverse futures is clearly crucial both to the de-sign of technology and the manner of its implementation. Despite growing global movements in favour of local development and circular economies (since at least the publication of *Small is beautiful* by Schumacher in 1973), the lure of large-scale technological infrastructure remains attractive to governments and corpo-rations who perceive problems and act primarily at this scale themselves (Scott 1998, Folch 2013).

Beyond extractivism

While it is appealing to critique rapacious forms of capitalism and colonialism, it remains crucial to understand how corporations and industries manage to con-tinue with practices that appear externally to be so destructive, and how they maintain support for their actions. Ethnographies like that of Jessica Smith (2019) take the reader into the world of coal-miners, to see the world from their perspective. Smith shows how the potential closure of coal-mining in

Wyoming is understood by miners as a failure on the part of the country to reciprocate the gift of warmth, light, and power that the miners offer by their labour, in providing fuel for power plants. The money they earn is a material benefit, but the additional costs in terms of their health and their contribution to what was framed as a patriotic duty to service the rest of the country was beyond remuneration. Closing the mines was thus a slap in the face of patriotic citizens, undermining the value of their contribution to the national good. This insight helps to explain the dedication of workers such as these to practices that environmental activists perceive as misguided to the point of being suicidal, or at least eco-cidal. Understanding this enables a shift in the practical steps for policy-makers who may wish to avoid confrontational resistance, by offering a different framing of the purpose of change, and a different horizon to which alternatives may be shaped. Vonderau and Müller's case study in this chapter demonstrates a similar point, showing how coal-mining embodies particular structural relations between civil society and the state (*pace* Mitchell 2009) that are reconfigured through the coal phase-out, generating new tensions and debates over post-coal futures.

Ethnographic research gives us a real insight into how people in different positions across institutions, systems or states, perceive the nature of change, and how they variously imagine, fear, or desire diverse futures. These may derive from the expectation of a reciprocal exchange (as in Smith's work), or they may point to a radical diversion from business as usual, as in the case of people in capitalist economies who choose to go 'off-grid' (Forde 2020, Vannini and Taggart 2014). Our aim in this chapter is to tease out how different visions of the future act in the present, primarily in relation to energy practices related to generation (rather than consumption).

Our case studies below take in the issues outlined above, ranging from the threatening future of coal closures in Germany, which we can contrast with Smith's work mentioned above, as a case of the dark side of the optimistic vision of renewable energy to which the European Union is committed. Protests about continued coal extraction meet another layer of concern as a village is threatened with demolition and relocation as the mine expands beneath its foundations. Environmental and conservation interests converge with government commitments to decarbonisation, but all are left with the troubling question of how to imagine a post-coal future. In light of the process and their understanding of what has already been decided, many citizens prefer not to engage in a tightly orchestrated 'future-vision' event – not only rejecting the bland and unrealistic visions being offered, but also demonstrating their lack of trust in planning as a governmental technology.

New renewable infrastructures also generate disquiet. As the case study below shows, Taiwan's policy for offshore wind energy brought energy futures into direct conflict with fishing activities, destabilising expectations of continuity in coastal life in ways remarkably reminiscent of oil and gas development. Protests against wind turbines are far from rare, appearing in countries from Mexico to Norway. We should not overlook the possibility that resistance is stoked by unwarranted fears (Oreskes and Conway 2010), not least after the activities of well-known ecologists in protesting against wind power in the UK in the 1990s, for example, based on debunked theories of risk. But it is also essential to acknowledge how different understandings come into being, what kinds of future visions are competing: between techno-idealisms, often large-scale or abstract, designed to address global concerns yet ignoring local contexts, colonialist extractive mentalities, concern for eco-systems and species, or objections to local environmental disruptions. It is not only interests that clash, but scales, timescales, and riskscales (Everts and Müller 2020) or temporalities of futures. Despite many years of neo-liberalization, citizens in Taiwan held to the normative vision that the state should provide services for the benefit of its citizens, and not provide profit for absent investors at the expense of local livelihoods.

In Mexico, national agendas around renewable energy are often misaligned with local politics, particular environments, and socio-political struggles. Changing the energy mix implies changing the political mix, and incumbents are resistant to letting go of the allegiances and dependencies that hold them in power, not least through clientelist networks. Solar futures that appear from the outside to be optimistic and future-oriented may appear unsettling and disruptive from the inside. NGOs promoting cheap and accessible solar energy tied to female emancipation experience resistance when they encounter moral worlds based in different principles and expectations.

Finally, to understand how the well-meaning problem-solving techniques of engineers and mathematicians can lead to less than ideal outcomes, a brief closer look at the techniques and limitations of engineering practice helps to show how promising technologies can be transformed through the political process into dubious generalisations and infrastructures that ignore local conditions. There is no need to demonise the actors or institutions that stand in the way of addressing climate change, since a closer look at their everyday experience of energy infrastructure reveals differences in motivation that appear reasonable. Better understanding of how energy futures emerge and appear to their beholders offers the prospect of dislodging acquired expectations. If different future visions can be appreciated, then different choices can be made now.

Throughout this chapter, and indeed throughout this book, there is an assumption that the future must be green (in the ecological sense) and we acknowledge that the notion of a green future has its own history, with changing moral and moralising assumptions in turn. An enduring dream of technological salvation threads through many current governmental and intergovernmental goals for decarbonisation and ecological modernity that do little to shift the economic structures that continue to enable extractive mentalities to endure (Lord and Rest 2021, Anand, Gupta, and Appel 2018). Much has been noted in the other chapters about the significance of thinking repeatedly about infrastructure, and this introduction brings the focus onto the development and governance of energy-related infrastructures, and their diverse modes of operation and various consequences. Throughout, we note the future ambitions that sustain infrastructural ambitions, those that challenge them and those that fall away.

The cases

The cases presented in this chapter draw on key issues in the future of and with energy generation of different kinds. Müller and Vonderau highlight the range of ambivalence around the prospective end of coal-mining in Germany's coal regions. From inside one of a series of anti-coal climate camps, they show how an anticipated energy transition is much more than a case of technological change, but emerges through distinct technologies of government. Who is in charge, who gets to participate in decisions about the coal-exit, and what is the significance of the large budgets allocated to compensate for the coal-exit? Why do offers of finance sometimes give rise to scepticism rather than trust, and how are promises for sustainable futures shaped by the history of energy visions that lived and died under the former GDR? Müller and Vonderau show how the politics of fuel translate into the reconfiguration of relations between civil society and the state, with moralised assumptions about a 'green' future and broad assumptions about the sustainability of future means of production.

Hsin-yi Lu demonstrates how far renewable energy falls into what could be described as more modernist visions of technological futures in Taiwan that are tied to regulatory withdrawal through development schemes. With wind energy presented as a *de facto* good, the destructive consequences of major infrastructure development on established fishing practices were de-emphasised, leaving local residents and fishing workers similarly sceptical about the 'green' credentials of wind power. Major development projects can so easily become controversial, revealing the kinds of moralisation embedded in energy discourses mentioned above. In what has been called a green-on-green conflict, Lu describes a kind of escalating morality-war between the 'cleanness' of green energy and the environmental and social justice due to the residents. The government increasingly struggled to control the agenda, as the sociotechnical imaginaries – developmentalist and the renewable energy vision – became increasingly stuck in a quagmire of conflicting future visions for 'good energy'.

Bresciani documents a different approach to changing energy regimes, with her study of Barefoot College's approach to developing solar power in Mexico. Its approach follows a broadly Gandhian, well-intentioned, positively gendered, South–South empowerment approach. However, it sits uncomfortably in a society where the political leadership and many others are primarily concerned with maintaining the status quo and focusing on the short term. Women who joined Barefoot College to enable them to take a role in installing and maintaining solar systems were not well received on their return, facing criticism and alienation for their attempts to better their position. In disrupting highly sensitive power structures, they found themselves facing a backlash that spread to the proposed solar power infrastructures themselves.

Looking more directly at the design and development of changing energy systems, Abram and Silvast report from the world of energy system modellers, mainly engineers, mathematicians, and economists attempting to evaluate changing technologies and their effects on each other at different geographical and temporal scales. Working in the relatively rarefied world of system-modelling, their ambition to create integrated models of whole energy systems can be seen to be ambitious and limited. The practical limitations of modelling are clearly recognised, but the challenge of coordinating models that are epistemologically different, embodying different conceptual approaches and varied ranges and kinds of data, presents ongoing difficulties. While all the models in question are pre-political, that is, they are mainly theoretical, they are imagined as informing future policy, whether investment decisions, regulatory frameworks, or operational schema. As such, the models themselves play a role in generating future options and choices, while concealing their own assumptions and normalised expectations for energy futures.

Future after coal: negotiating the coal-exit in Germany

Katja Müller and Asta Vonderau

Civil engagement to stop coal

In mid-May 2016, Proschim, a small village in Lusatia in the very eastern part of Germany, sees an unusual number of visitors. It is Pentecost weekend, but people have not gathered for any kind of religious event. They are here for the 6th Lusatian Climate and Energy Camp (*Lausitzcamp*). Proschim is threatened because the Welzow Süd lignite mine will be mined out soon and is planning a spatial extension. This extension will entail the resettlement of Proschim's 340 inhabitants, literally reducing the village to rubble. To prevent this, and to take a stand against climate change to which coal combustion contributes, people from near and far have gathered in Proschim. For a week, large circus tents have been set up right next to the village to house lectures, workshops, meetings and meals, and climate activists camp in countless small tents adjacent to it. During the previous week, about 1,000 people joined the camp, organising and participating in workshops, attending discussions, and making sure that everything from producing solar energy for the camp and communicating with public authorities to communal cooking runs smoothly. At the weekend, the number of camp visitors increases to almost 4,000 people.

On Friday, around noon, civil disobedience action starts. *Ende Gelände*, a loose alliance of people from many social movements and officially a guest at the Lausitzcamp, has planned and called for concerted action to block the Welzow Süd open cast lignite mine and the power plant Schwarze Pumpe. *Ende Gelände* uses a 'finger strategy', sending several groups of people towards the

---

Note: This case study is based on several research activities: (1) An ongoing research project 'Crafting Post-Coal Futures', conducted by Asta Vonderau. Focusing on the future-oriented forms of public participation and modes of knowledge, this project follows regional authorities' attempts to plan and manage coal-exit and its effects in Saxony-Anhalt, one of Germany's coal regions. (2) Six years ethnographic fieldwork for 'The coal rush and beyond' (www.coalrush.net) and 'Decarbonising electricity' (www.decarbenergy.net). Katja Müller has been conducting participant observation and interviews in Eastern Germany. (3) Results of the MA seminar 'After coal', led by Asta Vonderau (Anthropology) and Jonathan Everts (Human Geography) in winter 2019/2020. This entails walking interviews in ten coal communities in Saxony-Anhalt, investigating local inhabitants' relations to coal and their reactions to coal-exit.

planned blockades, who will act in concert (like a whole hand), yet separately from each other (as fingers) and able to react to and bypass barriers. This Friday, all 'fingers' reach the open cast mine and railway tracks unhampered; they don't even need to separate. The mining company Vattenfall stopped the diggers and production beforehand, and the guards and police have been considerate and restrained. People unfurl banners at a loading bridge and at the bottom of the mine, they sing, dance and celebrate. Some will stay overnight on the railway tracks and in the mine; other protesters will replace exhausted ones on Saturday. The protesters – overall about 2,500 – stay another night, and on Sunday noon the blockade ends, officially and peacefully. Only a handful of people with lock-ons on a digger and on the tracks stay on longer. They are later removed by the police.[1] Every group of protesters returning to the camp is greeted cheerfully and applauded, and as time goes on, the camp's feeling of exhaustion[2] but success increases. In the end, people block the connection between the power plant and the mine for 48 hours, and no coal is mined in Welzow Süd during this time. A civil protest march of about 1,500 people on Saturday afternoon comple-ments the protests, which make it to prime time TV news on Germany's public broadcasting network on Saturday evening. Furthermore, the protests form part of the worldwide *#breakfree2016* movement that sees action against coal-mining and fossil fuels in May 2016 all over the world.[3]

Ende Gelände at the *Lausitzcamp* 2016 was one highlight in a long series of anti-coal protests that has grown in Germany over the last ten years. The *Lausitz-camp* started as a small climate camp in 2011, with approximately 100 partici-pants in its early years. Bringing a long-term commitment to alliances with local people threatened by coal-mines, the organisers of the *Lausitzcamp*, includ-ing several from Berlin and Leipzig, successfully established bonds with affected individuals and civil society actors 'at the coalface'. The constant cooperation and support (Müller 2018), provided reassurance for continued protest on the ground despite setbacks,[4] contributing to an awareness of coal's relation to cli-mate change, and making sure that the range of protest supporters grew con-stantly in and beyond Lusatia.

---

1 Around 100 people tried to enter the power plant's premises forcefully on Saturday; the police stopped them. This is the only infringement of the 'consensus of action' that the protesters had agreed to among themselves previously, and also the only time police intervened with force.
2 Other issues, such as right-wing attacks on the camp and the town mayor's (faulty) legal no-tices to vacate the premises, keep organisers and participants busy.
3 https://breakfree2016.org/.
4 Small-scale protests, organized by locals, include protest marches and village celebrations stretched over time (see Müller 2018).

Fig. 3.1: Occupying a digger. Credit: Ende Gelaende CC BY-NC 2.0; 350.org/TimWagner.

There have been climate camps with an explicit focus on coal in the Rhenish coal-mining district since 2010, and in the vicinity of Leipzig between 2012 and 2014 and again since 2018. Climate camps are a concentrated format to educate people, to engage in networking about and beyond climate issues, to draw attention and voice protest through action and to transform climate ideals into lived practice for the duration of the camp (Feigenbaum, Frenzel, and McCurdy 2013). The German coal-focused climate camps are now a well-established element of the environmental movement that has grown stronger over the last ten years. They complement the work that environmental registers, associations, organisations, networks and action groups conduct, ranging from petitions, round-tables, and analyses to policy advice, symbolic protests, and mass demonstrations. The German Climate Alliance today counts 130 organisations representing 25 million (passive and active) members engaged in climate protection,[5] which in part overlaps with the looser action-oriented climate movement.

Civil society movements, and even civil disobedience, are important but not sufficient factors for moving beyond the coal rush (Goodman et al. 2020). They form an essential part of driving the exit from coal as part of the German Energy

---

5 https://www.klima-allianz.de/ueber-uns/das-buendnis/.

Fig. 3.2: Lusatian Climate Camp. Credit: Lothar Michael Peter (Beyond the Coal Rush www.coalrush.net).

Transition (*Energiewende*) (Müller and Morton 2016). Civil society movements have no decision-making competence and are not legislative bodies, but they inform decisions by voicing concerns and demands. While mining companies persistently promote narratives to argue that decisions to discontinue mining are purely economic,[6] continuous pressure from civil society is also a financial factor for mining companies and influences political decision-making. In Germany, 2018 saw the discussion and 2019 and 2020 legislative action regarding coal-mining.

Fig. 3.3: Pre-cut for lignite mining. Credit: Katja Müller.

Eastern Germany's coal-mining areas have more than a century's history of coal-mining and related industrial development (see Müller 2017). They were hit hard after 1990 by economic decline, an implosion of labour relations, and subsequent demographic decline. While 156,000 people worked in mining in 1989 in these two areas, their number had decreased to 20,000 by 2019.[7] As is well known, it was not only the coal-mining sector that was hit, but most spheres of society had to undergo some transformation (see for example Mau 2019, Hann 2002), a time experienced by many not just as structural change (*Struktur-*

---

6 https://www.lr-online.de/leag-aendert-tagebauplaene-39724043.html.
7 Statistik der Kohlenwirtschaft, https://kohlenstatistik.de/wp-content/uploads/2019/10/bk-ue berblick-1.xlsx.

wandel) but rather as a structural disruption (*Strukturbruch*). The problematic effects of Germany's reunification and the uneven relations between the country's east and west are still intensely debated in politics, in academia, and publicly. These contested narratives continue to impact contemporary ideas of citizenship, democratic processes, and attitudes towards the state.[8]

The decision on Germany's exit from coal

As a consequence of pressure from civil society, in 2018 the German Federal Government founded the Commission on Growth, Structural Change, and Employment (*Kommission für Wachstum, Strukturwandel und Beschäftigung*, colloquially called the Coal Commission), made up of representatives from politics, industry, academia, environmental organisations, and civil society. Its advice was to abandon all coal-mining and combustion for power production by 2038, thus effectively shutting down or converting Germany's 104 fully or partially coal-fired plants.[9] The Commission also recommended huge subsidies (€40bn) to be made available to support affected areas during the phasing out of coal-mining. These subsidies were intended to cushion the negative socio-economic effects of the structural change (*Strukturwandel*), anticipated in the coal regions, such as loss of jobs and demographic decline.

The government partly transformed the Coal Commission's advice into national law: In 2019, it passed the mining areas structural support law (*Strukturstärkungsgesetz Kohleregionen*); in January 2020, it passed the coal exit law (*Kohleausstiegsgesetz*). The legislative procedure was finalized in mid-2020. However, scientists and environmental NGOs (among them members of the Coal Commission), have criticised the coal-exit law as it does not correspond to the Coal Commission's advice for timely coal plant deactivation, and even allows for the activation of a new coal-fired power plant. The coal-exit law has therefore been criticized for breaking the consensus that the Coal Commission's diverse stakeholders had accomplished. The additional large 'compensation' sums for mining companies are also subject to critique.

---

8 This is arguably also mirrored in the rising popularity of right-wing parties.

9 According to https://www.bundesnetzagentur.de/DE/Sachgebiete/ElektrizitaetundGas/Unternehmen_Institutionen/Versorgungssicherheit/Erzeugungskapazitaeten/Kraftwerksliste/kraftwerksliste-node.html and https://www.umweltbundesamt.de/publikationen/daten-fakten-zu-braun-steinkohlen

Fig. 3.4: Lignite mine and power plant (East Germany). Credit: Lothar Michael Peter.

Managing structural change – making a post-coal future

On a late Friday afternoon in December 2019, a group of about sixty people, seated in a larger meeting hall, are waiting for the 'Future Workshop' (*Zukunfstwerstatt*) to begin. The event takes place in Halle, the largest city in the federal state of Saxony-Anhalt, one of the four coal-mining states that will be a major recipient of coal-exit subsidies. While claims and plans for distributing billions of euros have already been made by large companies and the politicians, Future Workshops, and many similar events are set up in different locations in coal-mining areas. In Saxony-Anhalt they are organized by the Innovation Region Middle Germany (*Innovationsregion Mitteldeutschland*), an association of cities and administrative districts from Saxony, Saxony-Anhalt, and Thuringia (which is the only state without coal-mining). The association aims proactively to shape the region's post-coal future, and to guide structural change (*Stukturwandel*) and its anticipated major economic and social changes. The Future Workshop is supposed to offer a forum for citizen participation in this process.

While some participants attend the *Zukunftswerkstatt* out of their own private interest, most of them have been selected and invited in advance by the organizers. The hall remains half empty, indicating that some invitees chose not to attend. Unimpressed by this outcome, a moderator welcomes those present and then quickly proceeds by bringing up the evening's key question: How would we like to live in Saxony-Anhalt in 2040? In order to formulate the answers, he encourages the participants to use yellow paper cards with pre-formulated statements and to pin these cards on a board according to their respective preferences. Cards contain statements such as 'In 2040, the region is a site for future factories', for instance, or 'The region is a site for research on new technologies in 2040', or 'Professional development will be essential for a fulfilled worklife'. The *Innovationsregion Mitteldeutschland* hence presents an agenda for what the future after coal will look like: markedly different from the present, better, more effective, and more sustainable. After little time for group discussion, the moderator sorts the statements into 'preferences' and 'non-preferences'.

Workshop participants wonder what this is all about. They are sceptical and do not really want to engage in this game of future-making. 'My guess is that all the coal-money has already been spent', says one of them, 'so what is our participation good for, really?' 'Why give an opinion about something that has already been answered by others?' another asks. 'All these statements are much too general anyway!' a woman intervenes. 'How could we possibly know what kind of "factories" they are thinking of? Will these factories be environmentally friendly, will they create jobs?' Finally, an elderly man raises his hand and summarizes the group's reaction: 'Why don't we put the preformulated statements aside and think of something entirely different, something which would make our region stand out?' A little overwhelmed by this request, the moderator provides the man with a blank sheet of paper and a pen and says: 'Sure, everyone is welcome to contribute.' The man writes his statement down and pins in the middle of the board. The statement reads: 'In 2040, there will be a huge nuclear waste repository in this region.' After four more hours of workshopping, the statement ends up among the 'non-preferences'.

Complexities and frictions of the coal phase-out

This intervention by distrustful citizens underlines the complexity and diversity of expectations, imaginaries, and consequences of the ongoing transformation to a post-coal era. It underlines that an exit from coal-mining and the concrete steps towards a coal phase-out are by no means accepted by all the affected actors. It is therefore unlikely that the exit from coal will turn out to become a self-evident or linear process.

In particular, the allocation of €40bn of financial support is subject to debate. This so-called 'coal money' has prompted various discussions such as that in the *Zukunftswerkstatt* as to how that money should be spent. As this example illustrates, Germany's civil society is asked to imagine a future according to prewritten statements; at the same time, leading (energy) industrialists met with heads of state and other politicians to discuss the industry's advances and needs at the *Ostdeutsche Energieforum*.[10] National and regional authorities are trying to promote coal regions as sites for technological innovation and expertise in order to claim funding for research and subsidies for industrial location (as has been the case in other European post-coal regions, and in line with ERDF future imaginaries). These authorities also stress the need to shape the post coal-future in accordance with local interests. As of now, a decision has been made on transport infrastructure only: to build a new motorway route across eastern Germany and to upgrade a railway line.[11] Local ideas are predominantly voiced at small workshops, discussions, and other initiated forums for public participation.

The mistrust and disruption at the *Zukunftswerkstatt* needs to be seen in this context, and in line with simultaneous talks between the Federal Government and mining companies about compensation to be paid for the power plants that have already been forced to close. Disrupting this arrangement prearranged by the organiser *Innovationsregion Mitteldeutschland* – a private company, funded by the state and federal government – signals disagreement with the form of public participation that had been programmed.

The scepticism expressed when negotiating the coal phase-out – especially in the context of eastern Germany – can be understood as a reflexive response to hegemonic narratives of necessary transitions, and broken promises of economic and social prosperity. In interviews[12] conducted in villages and towns around Saxony-Anhalt located close to coal-mines and power plants, both opponents and proponents of the coal-exit voiced scepticism regarding the official attempts to manage the transformation. Local residents state that they do not see how planned innovations will help solve everyday problems in their communities or be beneficial for them. For instance, where plans were mentioned to allocate funds for renovating the Romanesque cathedral in Naumburg, they were seen to benefit 'others', such as tourists, but not the people affected by the coal phase-out.

---

10 Participant observation by Müller in 2019.

11 The railway line between Berlin and Cottbus (where government bodies might settle) will be upgraded to Intercity level, thus shortening a little the time spent commuting.

12 Conducted in 2019 by students of the 'Nach der Kohle' MA seminar of Asta Vonderau and Jonathan Everts.

This complicated relation between state and citizens can in parts be traced back to collective experiences of inequality and deprivation (as noted above), becoming apparent in the scepticism and distrust about another 'manageable' change – the coal phase-out – that is officially declared as unavoidable, also expressed in a general sense of being tired of yet another need to change and adapt (Mau 2019). Fora like the *Zukunftswerkstatt* are perceived as turning locals into objects of economic and political experimentation, and transformation processes are seen to be initiated and designed by detached or 'external' expert groups.

Accordingly, when plans are proposed to fill the abandoned coal quarries in eastern Germany with rubble from West Germany, this bears the potential to reactivate past experiences and stereotypical images about the complicated relations between East and West. Such proposals display an economic and symbolic devaluation of localities and communities and serve as proof of continuous inequality between the country's regions. Similarly, Saxony-Anhalt's prime minister Reiner Haseloff protested against the government's suggestion to shut down an older power plant in his state to compensate for the rise of CO_2 emissions caused by a new plant, called Datteln 4, in North-Rhine Westphalia. He argued: 'People here do not understand why a new plant is going to be opened in the West, while workplaces in power plants and open cast mines will be lost in the East,[13]

However, beyond opening old wounds, the activation of Datteln 4 as part of the coal-exit law also serves to unify. It became a new target for regional, national, and internationally connected environmental protests. As the most tangible manifestation of breaking the consensus on exiting coal, it provided a site for environmental activists and civil society to voice demands for climate protection, speeding up an exit from fossil fuels and making a system change instead of climate change.

As we have shown, in the course of the German coal phase-out the former structures between civil society and the state are reconfigured. This triggers new controversies and inequalities as well as imaginaries and expectations. Environmental activists and sceptical citizens are important protagonists of this transformation, and they play a major role in negotiating Germany's post-coal futures. Although these groups are often described in the media and broader public as opposing each other – since they are differently organized, have different histories, and voice their opinions in different ways – they still have much in common. Both of their positions question official future scenarios and economic

---

13 *Die Zeit*, 15 January 2020, https://www.zeit.de/wirtschaft/2020-01/kohleausstieg-angela-merkel-kohlekommission-abschlussbericht-klimaschutz

models of prosperity and growth. They aim to alter a seemingly self-evident and linear economic and political development by confronting them with a diversity of actually existing and possible life worlds and futures (Pandian 2019, Rees 2018). These actors' relations to state authorities are based on scepticism and mistrust. In order to grasp the complexities of the coal-exit process, it is therefore necessary to acknowledge and investigate such different positions not simply as distrustful gestures, but as politically and ethically valuable forms of participation.

Wind futures: contested sociotechnical imaginaries of renewable energy in Taiwan

Hsin-yi Lu

Introduction

In the face of the global climate change crisis, greenhouse gas reduction has become an international imperative. All industrial countries are expected to be committed to low-carbon energy transition; Taiwan is no exception. The island's energy transition is further driven by the fact that 98% of the energy resources come from imported fossil fuels and uranium. To achieve the two goals of carbon reduction and energy independence, the government enacted the New Energy Policy on 3 November 2011, which promised to increase the ratio of renewable energy without compromising the principles of 'supply reliability' and 'price affordability'.[14] Under this policy framework, the 'Thousand Wind Turbines Project' was approved in February 2012, planning to set up 1,000 wind turbines with an aggregate installed capacity of 4.2 GW (now raised to 5.7 GW) by 2025.[15] Due to the scarcity of available onshore space, most of the future wind

---

Note: This case study is based on a research project 'Infrastructuring the Ocean: An Ethnographic Study of Taiwan's Offshore Wind Energy Development' funded by the Ministry of Science and Technology of Taiwan (2018–2020).

14 Bureau of Energy, Ministry of Economic Affairs, Taiwan, 'Policy of Promoting Renewable Energy and Current Status in Taiwan'.
15 Thousand Wind Turbines Office, https://www.twtpo.org.tw/eng/Home/, retrieved 27 February 2022.

turbines must be built in the uninhabited ocean if the policy target is to be ful-filled.

In the government's development scheme, the installation of offshore wind energy is divided into three phases: Demonstration Phase, Potential Site Phase, and Zonal Development Phase. Demonstration Phase began on 3 July 2012, when the Offshore Wind Power Demonstration Incentive Program was promulgated, which would subsidise up to 50% of installation expenditures for the industry to set up pioneer offshore wind farms. Three cases were selected in January 2013 as the awardees of the Incentive Program – the Formosa Demonstration Project, the Fuhai Demonstration Project, and the TPC Demonstration Project.[16] Potential Site Phase began in July 2015, when the Bureau of Energy released 36 marine zones deemed suitable for wind development. In the meantime, the government offers generous Feed-In Tariff (FiT) rates to attract multinational companies to enter Taiwan's offshore market. Seven companies were awarded with grid capacity to commission 10 offshore wind farms in 2018 with an estimated capacity of 5.073 GW by 2025.[17]

Since the change of regime in 2016, the incumbent president has made it explicit that offshore wind power is the best replacement for phasing out the country's three nuclear plants.[18] While the new regime and the wind industry are striving to turn the imaginary future into reality, however, the development of wind farms has encountered public opposition at all levels, as reflected in the voting results in local elections and the national 'Nuclear Green Referendum' in 2018.[19]

Below I use two cases that are three years apart to demonstrate the criticisms and challenges facing offshore wind power development at the local level in Taiwan. From these two cases, we see how the sociotechnical imaginaries (Jasanoff and Kim 2013) held by the state and the wind industry were received, interpreted, and contested by local people. In the first case, fishermen and residents near the development site appealed to environmental justice to challenge the legiti-

---

16 Formosa is located in Miaoli County; the other two Demonstration projects are located in Changhua County. The two counties are about 100 km apart.

17 For a most recent overview of Taiwan's offshore wind development process, see Gao et al. (2021).

18 Nuclear power currently makes up approximately 12.7% of the total electricity generation in Taiwan according to statistics provided by Taiwan Power Company https://reurl.cc/Ope4vD, retrieved 27 February 2022.

19 On 24 November 2018 voters overwhelmingly voted in favour of the opposition motion to continue with nuclear energy while alternative green energy sources were being developed, which means the government could no longer set 2025 as the nuclear-free deadline for the nation. https://www.taiwannews.com.tw/en/news/3584619, retrieved 16 June 2019.

macy of a specific offshore wind project. In the second case, another wind company framed offshore wind power differently in response to the above-mentioned environmental justice criticisms. But the vision that integrated environmental, economic, and procedural dimensions was still questioned by the local people. In the ensuing analysis, I will point out that Taiwan's existing power supply system was driven by a sociotechnical imaginary that can be characterised as 'high modernism' (Yang, Szerszynski, and Wynne 2018: 277), understood by the local people as 'the central government is fully responsible for the supply of stable, unlimited, and cheap energy'. This popular conceptualization of 'good energy' is obviously quite different from the liberal vision underlining renewable energy, and thus becomes the source of contention. That said, however, I wish to note that even though chasms exist between the national and the local energy imaginaries, the state and the industry actors still strive to adjust their energy discourses in response to the local doubts, and have gradually formulated a renewable energy imaginary that is closer to the local interests in aligning energy transition with rural development aspirations.

Research method

In August 2014 I started to examine the public controversies over the Formosa Demonstration Project, the first offshore wind farm project of Taiwan. I used in-depth interviews, focus group interviews, and participant observation to document the perceptions and opinions of fishermen and coastal residents about the Formosa project, as well as the social and environmental impacts of its development process.[20] Then in 2016 the government started to draft the 'Directions for Allocating Installed Capacity of Offshore Wind Potential Zones', which attracted several foreign wind developers to set up local offices in Taiwan to work on offshore wind development. I obtained research permission from one of the wind companies, Copenhagen Infrastructure Partners (CIP), which is a Danish fund management company that invests mainly in energy infrastructures. CIP has attained the EIA approval for Zhangfang and Xidao offshore wind farms, both located in Changhua county. With CIP's permission I began to follow its PR activities and community consultation meetings in Changhua county during 2018–2019. I recorded the conversations in the meetings and did follow-up interviews with local participants and CIP representatives. What I present below is

---

20 The results of this phase have been published in two Chinese papers.

taken from the research findings of the first and second stages of fieldwork, respectively.

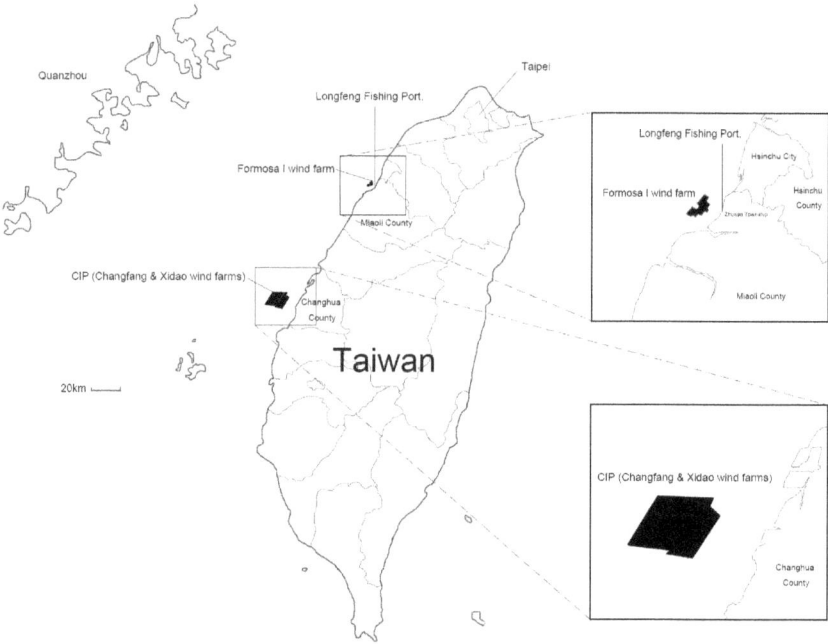

Fig. 3.5: Formosa and CIP OWF locations

Formosa Demonstration Project

On 1 May 2015, the Formosa project had already obtained a development permit, and the wind turbine construction was about to start. Thirty drift-gillnet fishing boats with white protest flags went out from Longfeng Fishing Port to the construction platform about 1 km offshore, at the intended site of the offshore wind farm. Later in November of the same year, Formosa's first wind turbine, also the first offshore wind turbine in Taiwan, was completed. Even though the wind turbine was not able to run electricity immediately, it received high praises from the authorities and the media. In just one week, the Minister of Economy and the Premier visited the Longfeng Fishing Port and praised Formosa as 'a milestone in the development of renewable energy in Taiwan'. On the day of the Premier's visit, there were two contrasting scenes in the small fishing port: one was the speech of the Premier, emphasising that offshore wind power could simultaneously drive the country's energy transition and green industry development. 'Taiwan has the best wind resources', he said; the country would become 'an exemplar to the world' in respect to its effort to transform the ocean into a major green energy production site. The other side of the ship's dock, however, was full of protest flags from fishermen, claiming that offshore wind turbines would affect their rights for fishing and survival. Although the fishermen's protests failed to block Formosa's construction, they clearly showed that the new energy technology had aroused suspicions about its potential environmental risks.

The fishermen's dislike of Formosa was actually representative of the general attitude of the residents of the coastal area. There are five wind farms and 54 wind turbines in the county, all located on the coast. The earliest set of wind turbines quickly became the target for public criticism; the subsequent three wind farms encountered strong opposition as early as in their planning stages. The main reasons for the protests were noise, landscape damage, ecological disturbances, and allegedly reduced production of shellfish farms from the large wind turbines. When the news of the Formosa Project became official, people resorted to their previous experiences of onshore turbines to apprehend the novel concept of offshore wind projects. In this area with densely installed onshore wind turbines, wind farms were akin to undesirable public facilities such as incinerators, landfills, and thermal power plants, which tend to be located in remote areas along the coast.

In the follow-up focus group interviews with the opponents, respondents also deployed environmental justice discourse (Schlosberg, 2004) to justify their protests. The first is the issue of distributive justice. Fishermen believed that offshore wind turbines invaded their fishing ground and interfered with the fishing rights of coastal fishermen. Coastal residents complained that they did not share the profit-making benefits of green electricity, but they had to bear most of the environmental costs from green electricity construction. Secondly, the issue of procedural justice. The interviewees believed that they had been excluded from Formosa's consultation process; even more, most of the cadres of the fishermen's association, who communicated most often with the wind developer, were not fishermen, and some of them did not even live in the coastal villages. Thirdly, recognition justice. Fishermen believed that the value of their way of living was not recognised by the government and wind power industry. Worse yet, nearshore fishing was accused of being the primary cause of damage to marine resources.

In addition to environmental justice, some opponents raised social justice issues. They claimed that they were not against green energy, but were driven by its violation of social equity as the state's FiT system used taxpayers' money to subsidise corporate energy providers, and the green energy facilities did not bring any benefits to the localities. Opponents complained that the ulterior motive of wind farm investors was to profit from the unreasonably high rate guaranteed by the government, rather than fulfilling environmentalist duties as the corporations claimed. A fisherman said in a loud voice, 'If it is for public welfare or national development, fishermen can make sacrifices...but green electricity is a scam. Its power generation is unstable, and it can only be bought by the government at a high price... Green energy is for private profit, not for national good.' Another hydropower technician who generated the power consumed in his household put forward his idea that large-scale wind turbines are expensive and can easily be monopolised by big corporations. Only small-scale solar power generators could achieve the goal of energy democracy.

The above anti-wind protest resembles what Warren et al. (2005: 853) term the 'green on green war': The wind industry employed the 'clean energy' discourse to justify its appropriation of marine space and fishing ground. Therefore the opponents must also resort to moral claims of environmental justice and social justice to challenge the environmentally friendly image of the wind power industry. Furthermore, what triggered the contestation over the desirable future of the area – whether it should be turned into a wind energy production base or remain in the status quo – was an agitated sense of continued marginalisation. Similar to what Franquesa (2018) observed in Spain, the fishermen's protest against the industrial-scale and corporate-monopolised offshore wind project was indeed a collective struggle to defend the dignity of their undervalued livelihood in the national energy development agenda.

CIP's Community Consultation Meetings

From the summer of 2018 to the spring of 2019, I participated in a series of community briefings held by CIP in each of the six townships along the coast of Changhua County. The meetings were set up to communicate the development ideal of offshore wind power and solicit suggestions from the residents as to the best ways to run their community benefit fund program. The formats of these meetings were similar: At the beginning, a CIP representative would use a well-designed PowerPoint to illustrate the scope of the project. Several themes have repeatedly appeared in these briefings: 'Taiwan's air pollution has become worse and worse because of the high ratio of thermal power generation, but now we have a better choice – offshore wind energy.' 'With offshore wind power added to the energy structure, we do not have to buy fuels from foreign countries.' CIP also emphasised that its offshore wind farms would create new job opportunities: 'According to the professional appraisal, 7,000 jobs can be created directly or indirectly linked to an offshore wind farm.' 'We have signed a contract with Chien Kuo University of Science and Technology [a nearby technical school] to train students. Our CEO will also teach a class, and the best students will be sent to Denmark for training.'

CIP also clarified the doubts from environmental justice and social justice perspectives. The representative said that the drilling technology and vessels introduced by Denmark will be used in the construction process to reduce the scope and extent of the impacts on the fishery. To address the social justice issues, the representative said, 'We are not a big corporation; we are a pension fund of common people. Also, we are the only foreign developer that invests in Taiwan life insurance. In fact, 30% of Taiwan's policyholders are also wind farm shareholders in a sense... So we don't strive to make quick and huge profits. Instead, what we pursue is long-term stable income for our shareholders.'"

From this series of briefings, it is clear that the energy sector of the government and the wind industry have articulated at this phase a new imaginary of renewable energy, which integrates global carbon reduction, energy independence, air quality improvement, domestic industrial development, rural economic revitalization, and human resource development all at once. It is not the same as the discourse constructed in the previous phase, which highlighted the 'cleanness' of renewable energy. This new energy sociotechnical imaginary seems to have effectively convinced a significant portion of the people, as I have never encountered in CIP's community briefings the same fierce public protests as in the earlier stage against the Formosa project. Nonetheless, there was still a gap between the public's interpretations and the messages that the government and the wind power industry intended to convey. One striking example relates to the sharing of economic benefits; most people regarded benefit-sharing as 'feedback money' or even handouts from a big corporation. In the first briefing session in the summer of 2018, the director of the county government's Green Energy Office told the audience, 'CIP has established a community benefit fund, providing 250 million NTD for the next seven years... You all can apply if you need it for education or research purposes.' Then several people raised their hands saying that the fund application procedure was too complicated, unlike Taiwan Power Company (TPC) and China Petroleum Company (CPC), both state-owned energy enterprises, who just gave feedback money directly to their neighbouring communities. In the November 2018 county election, the party that actively supported the development of offshore wind power was defeated. The newly elected county magistrate took the same position as her political party, which has questioned the cost-effectiveness of offshore wind farms in Changhua. This change of regime slowed down the county's offshore wind development until 2020, when

TPC completed its phase 1 installation of two offshore turbines. In a speech in June 2019, Chen Wenbin, former director of Changhua County Culture Bureau, explained the reasons for his party's defeat in 2018. Chen noted that the prospects promised by the government and wind energy developers 'were not internalised in the hearts of the people.' Most people were nonchalant about the energy future envisaged in offshore wind power development.

Yang, Szerszynski, and Wynne (2018) characterised the dominant sociotechnical imaginary in post-war Taiwan as a combination of 'high modernism' with national developmentalism. This imaginary has guided Taiwan's post-war energy policy, which was formulated to meet the ever-growing electricity demand accompanying the imagined perennial economic growth. In this imaginary, it is the state's responsibility to ensure stable, uninterrupted, inexpensive, and endless power supply through large-scale energy facilities and a centralised nationwide grid system. Furthermore, the state's construct of an ever-growing prospect in energy demand came hand in hand with the constructed possibility of 'power shortage' (2018: 28 – 29). The imagined lack of electricity raised fears among the people and strengthened their support for state-led large-scale power systems such as nuclear power plants, while renewable energy was said to be an alternative technology – immature, unstable, and unable to address the risk of power shortage. Even though the target set for offshore wind power generation is expected to exceed the current nuclear power capacity in 2025,[21] most people still regard renewable energy, especially offshore wind, as a dismissible source of power generation with an uncertain future. Since the people's established vision of 'good energy' has not been transformed, even those who support offshore wind power do so for reasons related to economic development, not from the ideal of a low-carbon energy revolution.

Conclusion

This case study presents two contested sociotechnical imaginaries underlining Taiwan's energy policies. One is the developmentalist imaginary, dominant in Taiwan's energy discourse throughout most of the postwar period. The other is the vision of renewable energy that emerged with global awareness of carbon reduction. My study documented how local people understood the two imaginaries

---

21 The incumbent president Tsai Ing-wen has committed to phasing out nuclear power in Taiwan by 2025. https://english.ey.gov.tw/Page/61BF20C3E89B856/e61c7f0b-9918-4c62-b80b-8a255f1f4aa8, retrieved 27 February 2022.

during the emerging period of offshore wind development. Between 2015 and 2018, there has been a notable shift in Taiwan's socio-technological imaginaries of green energy. In the former stage, green energy development was meant to keep up with the international trend of carbon reduction deemed by the then governing Kuomingtang (Chinese Nationalist Party) as auxiliary to the base-load supply of coal and nuclear. After the 2016 presidential election, power changed hands to the DPP (Democratic Progressive Party), and the official dis-course of energy transition began to shift. Expanding the capacity of renewable energy production is not only desirable for environmental reasons, but also aligns with the new agenda of economic policy that emphasises innovation and sustainability. The CIP case illustrates how the European developer has at-tempted to promulgate a new imaginary for renewable energy, one that conforms to the government's green economic agenda by integrating offshore wind gener-ation with local economic improvements (see also Chien 2020). The company's efforts were frustrated as Taiwan's energy market has not been fully liberalised; all the energy producers sell their electricity at a generous FiT rate to Taipower, which then distributes the electricity through the national grid. Therefore the so-cial expectations of the state being responsible for maintaining a stable and ac-cessible energy supply remains intact against the backdrop of global neoliberal-ization. While this way of evaluating different energy systems might seem future-reverse, I wish to point out that it actually gives clues as to what people imagine the just energy future should be – that is, a fair right to energy as integral to cit-izenship and guaranteed by the state.

Under a rising sun? Solar energy, bright prospects, and missed futures

Chiara Bresciani

A Gandhian solar utopia

The paradigm of energy transition frames renewable energies as intrinsically de-sirable, and solar energy in particular has been conceptualised as unquestiona-

Acknowledgement: The author is most grateful to the women who opened their homes and shared their stories, memories, accomplishments, and concerns with me. As Solar Engineers

bly good, its proponents arguing that locally sourced solar energy could generate new political scenarios by emancipating citizens from a central authority powered by electric grids (Scheer 1994, 2004). Such promises of free access to energy-intensive modernity have helped to establish links between solar energy and international development projects (Cross 2012). The case presented here concerns one such solar electrification projects, managed by the Indian NGO Barefoot College in a remote community in Southern Mexico.

Barefoot College has been among the first to seek to advance economic and social development of marginalized and rural villages through the installation of domestic solar panel systems. Barefoot College was founded in 1972 in the village of Tilonia, Rajasthan (India) by educator and social activist Sanjit 'Bunker' Roy, with the aim of providing simple and sustainable solutions to the problems of rural India. Since 1989, it has been running community electrification initiatives targeting off-grid, rural villages. This has since become its main sector of activity, with the others following, e. g. night schools were open after solar lanterns made it possible to light up the classrooms. To Barefoot College, solar energy is a way to alleviate poverty and a tool to accomplish long-term political objectives of empowerment and development. Solar kits are thus little development devices (Collier et al. 2017) that can have life-changing effects for individuals and for entire communities, and this can only be achieved with a focus on rural communities themselves, and women in particular. The objective of Barefoot College is to transform villages into self-reliant, autonomous, and attractive places to live by ensuring that members of rural communities are in control of the technology and resources.

It is in its focus on self-reliance and autonomy that Barefoot College's Gandhian ideological roots are most evident. Gandhian philosophy advocated for national independence through labour, believing that economic self-reliance would help sever imperial bonds with the colonial ruler, foster rural autonomy, and improve the lives of the poor (Gandhi and Jack 1994). Following the Gandhian spirit of equality across classes, castes, and genders, Barefoot College considers women to be key to social change and privileged agents of community development, and places them at the centre of its electrification projects. The first step of a solar electrification project is to visit the target community and select three or

of Barefoot College they are trained to be a source of change and inspiration for their communities: to me, they are above all an unparalleled model of courage, strength, dignity and dedication. I am also indebted to Gubidxa Guerrero from the Comité Melendre, Juchitán for his time and kind assistance in the early stages of this research. A special acknowledgement goes to Rodrigo París, Head of Latin American and Caribbean operations at Barefoot College, for his generosity, support, and openness to discuss the results of this investigation.

Fig. 3.6: A Barefoot solar kit constituted by a battery, a lantern and three lamps that can be hung up. Credit: © Chiara Bresciani and taken in Cachimbo in 2015.

four grandmothers between 36 and 55 years old, illiterate or semi-literate. The selected women from rural communities mostly from across Africa, South-East Asia, and the Pacific are brought to the Barefoot College campus in India, where they are trained for six months in the assemblage and installation of solar kits. Upon completion of the training, the women become BSEs (Barefoot Solar Engineers), return to their communities and start installing solar kits whose components are shipped to them from India. From that moment on, they will also be responsible for their maintenance, receiving a part of the subscription paid by the households who benefit from the project. The choice to invest in older women is motivated by a perceived need to ensure that they will not seek paid employment elsewhere but will instead remain in the villages and pass down their technical skills. The expected outcome is to raise their status in the community and turn them into role models for the other women, based on the assumption that social progress will cascade through society from the implementation of off-grid solar electrification projects. This case study looks at how this particular form of solar utopia has played out in the remote island of Cachimbo, at the border between the States of Oaxaca and Chiapas, Mexico.

Fig. 3.7: A Solar Engineer fixing a solar lamp for a family. Credit: © Chiara Bresciani and taken in Cachimbo in 2015.

Missed futures of solar energy generation

Following the strategy of the Indian government (cofinancer of Barefoot College) to further its engagement in South–South cooperation, Cachimbo was chosen by Barefoot College as its first project in Latin America, the last continent in its successful expansion outside of Asia. The choice of Cachimbo was based on its relative isolation, its poverty, and the devastation caused by Hurricane Barbara in 2013. The resulting emigration has led to the progressive abandonment of the community, which is in danger of disappearing. Although politically dependent on the municipal authority of San Francisco Ixhuatán on the mainland, Cachimbo island remains off-grid. In 2014–2015 I was conducting fieldwork in San Dionisio del Mar, a Huave village in the predominantly indigenous region of the Isthmus of Tehuantepec, Oaxaca. My main interest was the small-scale social and cultural impact on the community of the proposed development of a wind energy park. In the last ten years, the wind parks in the Isthmus have been at the centre of violent protests with international media attention and have been a privileged spot for anthropologists interested in the entanglements of renewable energy, green extractivism, and indigenous movements (Dunlap 2018a, Dunlap 2018b, Howe, Boyer, and Barrera Pineda 2015). Within a regional context marked by tensions, factionalism, and violence, enthusiastic reports on local media about the solar energy project in nearby Cachimbo seemed a much-needed success story of renewable energies about to fulfil promises of prosperity and peace in a remote, off-grid community. After securing access to the field site through the local organisation supporting Barefoot College, I visited the island to conduct ethnographic fieldwork. My aim was to collect insights into the effects of community-managed renewable energy projects on local societies and politics, rather than the analysis of changes in electricity use and modes of consumption. I initially spent one week in the community, hosted by the one of the Solar Engineers, conducting interviews with three women participants, local authorities, and beneficiaries and opponents of the solar project. I also engaged in participant observation of the women's daily interactions and their domestic life and work, both in the repair workshop and the home visits to fix solar equipment. Two months later, I took part in a round of public events in the State capital of Oaxaca on renewable energies, which included presentations and talks by Barefoot College representatives and the women. Later on, I made a follow-up visit in Cachimbo and conducted interviews with the coordinator of the local organisation and the head of Latin American operations at Barefoot College.

At the time of the research in June 2015, around 44 solar panels had been installed, 33 in the main settlement and in the ranchos of the island and 10 on the mainland. Most solar kits were still in boxes stored in the workshop of the Solar Engineers. In fact, upon the return of the women from India in March, many people who had initially declared they were in favour of the project decided not to subscribe to the service, with others returning their kits in the following weeks. The reason most commonly given to me was the monthly cost of a solar kit of 150 pesos/month, which they deemed too high, although it had been agreed by the inhabitants of Cachimbo during the first visit of Bunker Roy. This amount is higher than the cost of the battery-powered torches and the oil lamps used at night, and allegedly higher than what is paid on the mainland. People surveyed in Cachimbo reported that the average cost for electricity sustained by their friends and families in town was 80–100 pesos for two months, enough to cover light, TV, refrigerator, possibly a washing machine, and, at a slightly higher cost, also a freezer. The solar kit was thus deemed to be too expensive, and the energy provided insufficient for their needs. It is worth mentioning that in the rural fishing communities along the coast of Southern Mexico, ownership of refrigerators and freezers is at the same time a sign and a source of relative wealth, as they permit the storage of a small surplus of fish. Freezers in particular can constitute an additional source of income through the sale of bags of ice to the fishermen who need to keep their product fresh until they are able to sell it.

Another theme emerged during fieldwork: the critiques, the gossip, and the ostracisation of the women by many of their fellow villagers. The beginning of

Fig. 3.8: Interior of a house with a bulb powered by a solar kit. Credit © Chiara Bresciani and taken in Cachimbo in 2015.

this animosity was traced back to the day of the women's return from India, when the boat carrying them to the island was met by the boat of the Presidente Municipal (the mayor). The women, escorted by the leader of the local NGO who had been supporting and organising their travel to India, rejected the presence of the Presidente in the small welcome party organised by the villagers, which resulted in his 'public humiliation' and departure. According to the NGO, the local authority had tried to hijack and sabotage the project since the beginning, trying to impose women close to the mayor's entourage as participants and later delaying the issuing of the documents needed for the trip to India. The women and the representative of the NGO objected to the mayor taking credit for the project, and, according to my informants, that was when 'the conflict started and the community ceased to be united'. Most people expressed shame and concern for rejecting the mayor, whose support they see as essential for the survival of the community. Hostile and concerned, many people opted out of the project, and the owner of the building where the solar engineers had established their workshop refused to lend it for the following year. Many people, including some tied to the women through relations of *padrinazgo* (a sort of ritual kinship among Catholics), stopped talking to them or buying food from them, describing them to me as presumptuous and cocky. At the same time, their husbands were criticised and mocked by other men for letting their wives travel alone outside of the island, which led to an increase in intra-family conflict. While the women insisted that they had travelled to India to the benefit of their community, the villagers complained of their 'selfish and irresponsible' behaviour and responded with increased social control and stigma. The project basically came to a halt in the community and at the moment of fieldwork its future was uncertain. At present (2022), electrification work is being carried out occasionally and only on the mainland.

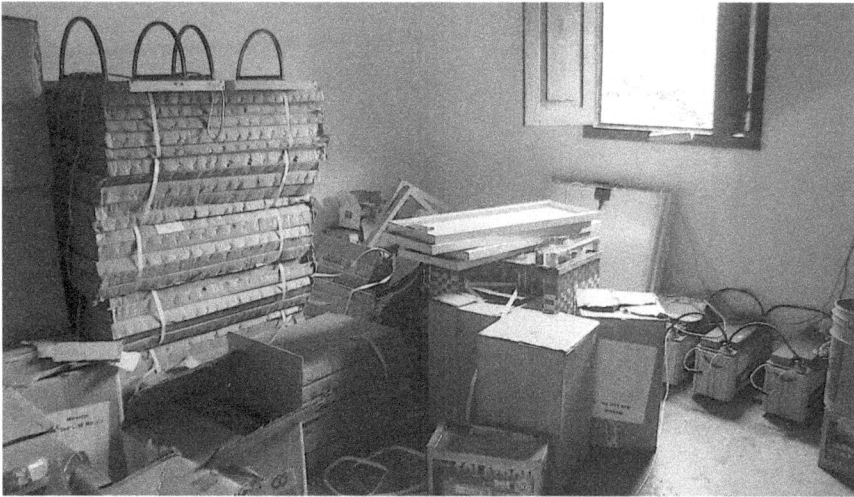

Fig. 3.9: Left-over solar kits in the workshop. Credit © Chiara Bresciani and taken in Cachimbo in 2015.

Expectations of energy consumptions and the backlash of the status quo

There are two explanations for the failure of the project in Cachimbo, the first concerning energy expectations and imaginaries. The solar lamp officially became a benchmark for the minimum level of sustainable energy after the UN's quantification of the universal basic need for electricity (Cross 2017). In line with this, the electrification goal of Barefoot College is to provide lanterns and small solar panels for domestic use, suitable for illumination and recharging of mobile phones. To the project participants, a kind of 'energy frugality' is proposed as the solution to energy poverty. A central issue thus emerges around the definition of how much is 'enough', and who is to judge that. The inhabitants of Cachimbo are somehow at odds with the basic needs attributed to them and in the interviews they repeatedly bring up the inability to power energy-intensive white goods and domestic appliances such as fridges, TVs, and stereos with the solar kit provided. Frugality is not part of the prevalent value system of the region, possibly because of its association with poverty: as for many people in middle-income countries, getting out of poverty means joining the middle class and adopting its consumption models and status symbols. The case of Cachimbo warns against the consequences of a 'one size fits all' approach that disregards social and cultural specificities in favour of a universalistic ideal of development (Collier et al. 2017).

The geographic location of Cachimbo posits challenges different from those previously encountered in locations where Barefoot College had built its expertise, such as rural Africa or India. Despite its remoteness, Cachimbo's economy and society are in fact deeply entangled with the municipal centre through kinship networks, economic and institutional dependence, and regular visits to its market. While rural settlements in Southern Mexico are often characterised by extreme poverty, Mexican society (as much of Latin America's) is epitomised by a high inequality rate rather than by generalised poverty. Through migration, and in the case of Cachimbo also due to the proximity to the municipal centre, the possibilities offered by consumer society are well known. In Mexico, goods manufactured by global brands are widespread in even the most remote settlements. Consumer imaginaries are constructed through media but also movement and participation of people in regional trade, and while this is hardly enough to move people out of extreme poverty, cash flows through remittances and state programs for income support. A first reason for the failure of the project can thus be ascribed to the incompatibility (or better said: misalignment of expectations) between proponents and beneficiaries of the project with regards to the quality of the energy provided. As Barefoot College projects are based on sustainability and aimed at bringing light to people living with less than $1/day, there seems to be a fundamental gap between the way the project was conceived and people's imaginaries.

A second issue concerns the ways in which people imagine energy futures within pre-existing political and economic structures. The political and social landscape in Southern Mexico is organised around clientelistic networks in a peculiar political model known as *caciquismo* (Nuijten 2003). The case of Cachimbo highlights the fissures in the interplay of the energopowers exercised by different social actors: while the Municipal President aims at maintaining and acquiring political control through infrastructures, Barefoot College's explicit goal is for

Fig. 3.10: Solar panel on the roof of a house in Cachimbo. Credit © Chiara Bresciani and taken in Cachimbo in 2015.

rural, disadvantaged communities to gain self-sufficiency. Meanwhile, most people's ambition seems to be gaining full citizenship through their insertion in the clientelist networks. As the manager of the local partner NGO has pointed out, the relation between Cachimbo and its municipal capital is of abandonment but also of dependence: despite its absence (lack of public services and scarce attention to the village's needs in the aftermath of the hurricane), the political authority is afraid that the clientelistic networks that tie it to the community may be severed. Within a political landscape widely managed through patronage, where electoral fraud, vote-buying, and corruption are common, the loosening of clientelistic ties would mean the loss of votes for the municipal authority, and the loss of occasional support for the villagers who promised their allegiance by selling their votes. With the solar electrification project, this relation of energetic dependence is being broken, as advocated by those solar utopians who linked freedom to photovoltaic technology. However, the municipal authority is well aware that at this point, the dependence may also be interrupted in other areas: a poignant example is that of the Comité Solar, a civil association legally constituted out of the collective agreement between the solar engineers and the project beneficiaries. Its legal status allows its members to discuss their necessities and negotiate directly with state and federal authorities without having to go through the municipal presidency first, a possibility whose importance becomes clear when considering how the municipal presidency had actively countered the issuing of travel visas to the women.

Most people in Cachimbo did not intend to challenge the status quo, and supported the project only as long as the existing political and social (including family) structures were preserved. Ideals of development, emancipation, and self-sufficiency of rural communities collided with the existing relations of dependence and clientelism. Little development devices dream of being off-grid, but "many turn out to rely on material, administrative, and political infrastructures [...] on which they remain dependent" (Collier et al. 2017). Sometimes they just do not seem to be working, since in practice their ambition is, in fact, 'to bypass the great institutions of government, business, education and religion [...]. This more aptly describes the ethos of those who design little development devices than of all their intended users' (Collier et al. ibid.).

Conclusion

The failure of the solar electrification project in Cachimbo has been caused by the incompatibility between two models of political action that dramatically clashed under the specific circumstances created by the project. Each of them pursued a different future: Barefoot College and the supporting NGOs aimed at a radical future characterised by political change and social progress, and expected the villagers to be keen to gain self-sufficiency and independence from the municipality, and also grateful and supportive of their women engineers. However, local authorities and most people in Cachimbo had only hoped to improve their living conditions while preserving the status quo. Familiarity with the social context and ethnographic methodology might have identified such misalignment of expectations and socio-technical imaginaries, and showed the different ways that people thought of as "beneficiaries" conceive desirable energy futures and the place of energy in their ethical worlds.

Fig. 3.11: Two Solar Engineers on their way to check and repair solar equipment across the island. Credit © Chiara Bresciani and taken in Cachimbo in 2015.

Modelling the future?

Simone Abram and Antti Silvast

Introduction

What are research engineers doing when they develop models of the energy system? While modellers may or may not be designing specific technological interventions, their modelling activities play a crucial role in the progress of technological ideas and imaginaries. Contrary to the popular vision of inventors tinkering in workshops – an image promoted by appliance developers such as Trevor Bayliss – the majority of significant engineering and infrastructure developments come out of research laboratories through a lengthy process of detailed modelling. What is meant by modelling can vary extraordinary widely. Models may be calculative computing models of inputs and outputs or processes, they may be based on readings from real physical entities, simulated data generated mathematically from physical measurements, or actual material exemplars – even a whole house fitted with monitors that acts as a 'model' or archetype of material effects. Common to all such models is the measurement of something in the real world which produces mathematical data that can then be manipulated with the use of computing software, using programmes often designed and run by those researchers who we refer to, in shorthand, as modellers.[22]

Envisaging a whole system future

The laboratory is a large open-plan office space in a large northern university known for its engineering excellence. Hidden in a middle floor of a multi-storey building, the long rectangular space has windows onto an internal courtyard along one side, strip lighting along the ceiling. Pairs of L-shaped desks with low separators sit in serried ranks, each with a black wheeled office chair, computer screens, stacks of papers and books, and a motley collection of pens, books, calendars, files, jars of coffee, bags, jackets and even coffee machines congesting the desks, cluttered with mugs, coffee jugs, plastic water bottles, lamps, rulers, university telephones. Modellers working in the office come from many different countries to do doctoral research here, and are of diverse ages from early 20s to mid-40s, but all have science degrees and in this office they mostly have engineering or maths degrees, some economics. Under a third are women, which is a relatively high proportion in this context.

What is it that is going on in these offices? According to Antti Silvast's informants, those working here are engaged in diverse forms of modelling with a common aim, to better understand an energy system in transition – i.e. a 'future' energy system that is confronted by the impacts of climate change.

In technical terms, they are mostly using coding packages such as Matlab, or Simulink, coding or recoding for new applications. Some models are economic models, some are physics or mechanical models, some are design-based; all aim to achieve a correspondence between the model and reality, at least to some degree.[23] The challenge for all the ESI modellers is to work out how

---

22 This case is based on empirical research carried out under the auspices of the National Centre for Energy Systems Integration (CESI), a 5-year research centre project funded by the Engineering and Physical Sciences Research Council (EPSRC) UK, and cofunded by industrial partners including Siemens and others. Our role in the project was to conduct an analysis of the energy systems modelling process itself, and to contribute social science knowledge to the integration process. Antti Silvast spent a term in university modelling laboratories, interviewing around 30 engineering modellers, some of whom are directly engaged in the ESI project, others who are conducting PhDs in engineering on other, related topics; Simone Abram interviewed various CESI investigators and participated in numerous project research meetings
23 Proprietary energy models such as TIMES, MARKAL, ARIMA, MonteCarlo, are rarely available to doctoral students, although some researchers have access to these systems.

Fig. 3.12: Office. Credit: Abram

their different models can be linked together, in order to take a 'whole-systems' approach to energy infrastructure investment, operation, and management. The aim to analyse and understand a whole system is in some senses very similar to the anthropological ambition to make holistic analyses. While for an anthropologist, that might suggest incorporating historical depth, taking a multi-species approach, or considering individual practices in broader cultural or social contexts, the ambitions of the energy modellers is more modest – to see, for example, how a gas system can interact with an electrical system, or to evaluate how one type of fuel (such as hydrogen) might offer services to a range of interconnecting systems of infrastructure (such as fuel for transport, or energy storage between electrical and gas systems). The whole-system approach poses similar challenges, however, in requiring 'cuts to the network' to be imposed, in order to frame any particular question (Strathern 1996). For the modellers, a whole system model of everything was a fascinating challenge that rapidly emerged as an impossible one without tighter restrictions on what might be included at any one time.

After almost a year of discussion about how to combine diverse forms of model, the group decided to scale their attempts, that is, to start with a whole model of a small, clearly limited location, then if that were to work, to progress to a medium-scale, more complex location, before progressing to national scale models. The choice of location for the first iteration of the model was decided after much discussion on the basis that one of the research groups had worked for some years with a small intentional settlement in Scotland, and had extensive data on their energy use, including ongoing monitoring of energy flows. It was the availability of data that primarily drove the choice. The first stage of the model-building included gathering of relevant data, including a detailed map of all the buildings, a schematic map of the electrical system of the settlement, showing all the electrical equipment, demand data for different kinds of dwellings on site, and survey research on energy practices among residents.

Two issues presented difficulties in modelling the whole energy system for this settlement. The first was that the settlement had no gas supply, making it untypical of the kind of general model the group intended. That could be solved by producing a replica of a gas system that would be typical of this scale of settlement in any other part of the UK. The second difficulty was that earlier empirical research revealed that many of the houses had wood-stoves, and residents largely sourced their fuel informally. Primarily, they retrieved excess forestry products from a nearby plantation with whom they had a non-financial agreement (i.e., they scavenged wood for free). This wood was neither measured nor accounted for, emerging as it did from outside the formal economy. The only means of estimating its contribution to the energy system was by extrapolating it from any differences between recorded energy use and building heat records. As this was unreliable, the modellers chose instead to ignore this source of heat. Resulting models would therefore assume that no wood-fuel was used, despite its significance in everyday life.

Whilst the modellers recognised that these two problems moved the resulting models some way from lived experience, when challenged they tended to shrug off the discrepancy as one that they could not incorporate. Even though this meant that no future models would incorporate wood-stove-heat as an element in the future energy system, they were prepared to consider this to be not sufficiently significant as to undermine their system models. The modelling exercise continued, identifying a baseline model based on thermal archetypes (for buildings), standard technical electrical and gas network models, and storage, before projecting future conditions in relation to diverse climate, technology, or energy supply scenarios. The resulting model would focus on the impact of:

i. behaviour (in relation to, e.g., demand system modelling)
ii. transport and electric vehicles
iii. new forms of onsite and community energy storage

From our perspective, however, as ethnographers, these compromises flagged up future problems that might emerge from the framing of the whole system. It was not only that wood fuel disappeared from the model, but that wood fuel therefore disappeared from any future projections for energy systems in the UK, while gas remained. In other words, political and moral choices had been made by dint of being practical solutions to modelling difficulties with access to suitable data (Parker and Winsberg 2018, Rudner 1953). Clearly the model does not represent a future of or for this community, nor does it represent the present. Instead, the model represents a possible future model of integrated energy systems that could become the baseline for modelling the UK's regional and national energy worlds, on which decisions about future developments could be built.

Clearly, there were considerable uncertainties about how closely this model could be said to represent anything, not least the settlement being modelled. Questions about uncertainty are often translated into debates about validation. If a model represents reality, it necessarily only represents a selection of that reality, and it simplifies. A model is not a Borgesian one-to-one replication, but a codification, and thus the process of developing a model entails a long series of choices and approximations. Which factors are important to include, which can be set aside, how accurately must each factor be reported: these questions are implicit in the design and development of models, such that the limits of a model remind us of the limits to defining an ethnographic field. Suffice it to say here that models are normally iterated, to meet levels of accuracy that are defined in relation to purpose. Validation is sometimes complemented by what

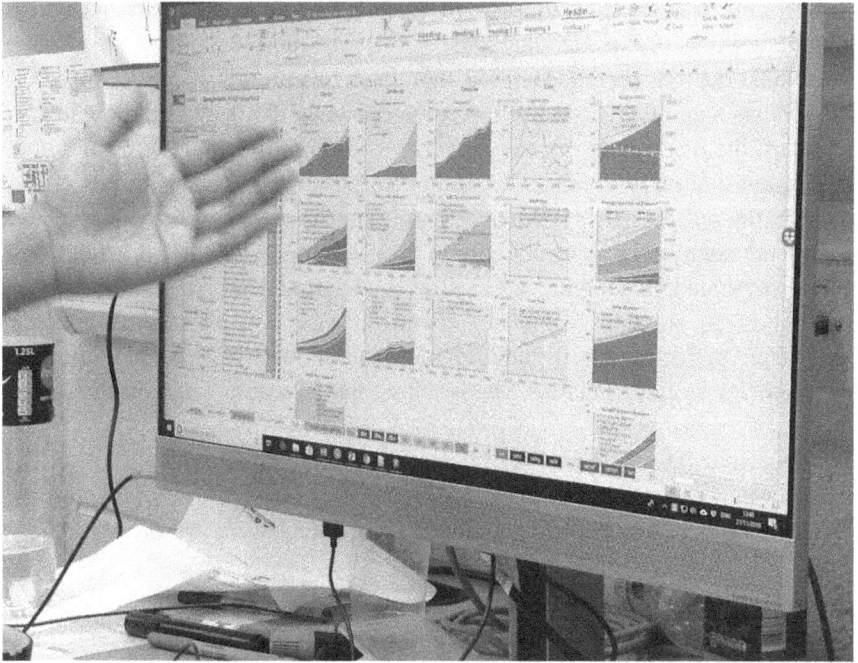

Fig. 3.13: Digital models. Credit: Abram

is known as 'expert elicitation' – i.e. asking people defined by the modellers as expert (whether technical expert or socially informed) to evaluate how effective the model is and whether it is good enough for the purpose envisioned. As Parker and others show (Parker 2014, Peterson and Zwart 2018), levels of accuracy can become embedded in models that are then incorporated into other models, solidifying the errors (or 'uncertainties') from one model into another. Alongside these limitations, boundaries and tolerances, are a set of presumptions and assumptions about the future.

Futures, projections and predictions

Models incorporate a range of future-oriented concepts and forms of future. Some models are replications of the present that can use measurements from sensors in the current world to generate predictive processes for future conditions. In a simple example, sensors that measure the energy consumption of metro trains, combined with models of solar radiance, can be used to predict energy consumption in alternative configurations of metro trains, or future consumption as weather or climate vary. Such futures can be conceptualised as a variance of the present, if we argue that all other contextual factors are assumed to remain constant. Similarly, models of train passenger demand may take current conditions and vary some factors to produce the future not as an input to the model but as an output – the model shows what may happen if certain factors vary. In other words, these models explore and test present assumptions, in what might be thought of as a future where such assumptions are already in place, or what we might call 'present futures'. Such futures have in common with other concepts of future that they project the present forward, allowing for some selected elements to change, but others to remain constant. In the type of energy system model discussed here, the elements that remain constant tend to be the larger, more socially oriented contextual factors such as the political or economic structures. Energy use and production models may assume that flows of fuel may vary, but they rarely model for a revolutionary change in market structures, for example.

These futures should not be confused with *future scenarios* of the kind produced by National Grid. These scenarios are elaborated around a two-by-two matrix, using models such as UK Times to project possible outcomes based on different rates of change along the axes promoted (Loulou et al. 2016). These are clearly politicised and partially contextualised projections, feeding regulatory and political ambitions through a technical and economic modelling evaluation to outline investment priorities. For example, the National Grid's axes changed from prosperity/green ambition (National Grid 2017: 10) to speed of decarbonisation/level of decentralisation to meet 2050 carbon targets as 'the relationship between green ambition and prosperity has changed' (2018: 14).

The well-established TIMES model uses external data on population, GDP, households and so forth, and is, as such, vulnerable to all the inadequacies and flaws in such data. It further insists that such data generate coherent output growth rates in different regions, again suggesting deeply ideological assumptions. Research on planning for housing demonstrated unequivocally that the only agreed understanding of population figures is that they are wrong, yet because they exist, they are used to guide housing development policy at national, regional, and local levels.[24] In engineering terms, one might say that population figures are good enough – indeed engineers describe their own future projection activities in terms of being 'good quality' rather than 'accurate'. Clearly, a model cannot be validated against the future, since future observations are not available. Instead, as models appear to align with real data – i.e. actual observations –

---

24 See Murdoch and Abram 2002.

Fig. 3.14: [No caption] Credit: Abram

the modeller can become more confident in their ability to model diverse future conditions. In this future, the modeller's own competence has also increased as the model has become closer to reality, so the direction of travel is one in which both the model and the modeller's abilities become more closely aligned with anticipated future observations and the model more and more accurately reflects the observed world. But, of course, the future is also continually receding, and therefore the modelling process remains ongoing. Models have to be updated to adapt to a changing reality in the present, and changing expectations of the future. For example, current energy system models have to take account of new developments in energy storage and radical changes in energy costs such as the sudden drop in UK wind energy spot prices in 2020 by nearly half, which changes the decision-making algorithms for future energy planning. Current debates about the emergence of electric vehicles, and consequently the changing demand for EV charging capacity, also loom very large for energy-demand modellers, for example, and they think a great deal about such issues, wanting to know how they would affect the calculations in their models (see Ch. 2).

Models themselves therefore also have futures as well as pasts, and the future of a meticulously developed model may also be deeply uncertain. Modellers agree that scenarios for the future are inherently biased, and are affected by vested interests. Rather like social scientists, they are aware to some extent of their own interested roles, and attempt to make these explicit where they can – yet they less commonly have the sorts of social science techniques available to them to do this systematically, or to appreciate the depth of cultural assumption and positioned perspective that reflexive theories offer.

Summary

When discussing emerging technology, it may be tempting to look at product design or laboratories, but we argue that infrastructural technologies more commonly emerge through stages that include modelling and simulation processes. While one might assume that models are future-oriented, what future-orientation means can incorporate many different kinds of future, and many different orientations. Some of these future orientations are possible futures, some are impossible, and others are implausible. The evaluation of what is implausible or what is likely can be seen as a weak link in the modelling process, precisely at the point where the social and technical are most closely aligned, in the political, regulatory, and real economic practices that people engage in. While the future behaviour or performance of motors or cables may be relatively stable and accessible to simulation, their performance in the real world of climate change, geopolitical instability, changing household formation, and so on are immediately more challenging for modelling approaches. Attempts to integrate models of different sectors of the energy system carry with them dangers, not only of the potential failure of an integrated model, but through the carrying-through of errors and assumptions that may result in injustices in the longer term, in relation to infrastructure decisions, investment strategies, or changes in relative energy costs between technologies, between regions, or between social classes.

Looking through our notes from interviews with modellers, we find over 40 different ways of discussing the kind of future that models engage with. What is emerging is not only a technology but a range of possible futures that open up and close down at least partly as a result of how they move through the modelling sequence. Some anticipate future demand and explore how infrastructures could meet it, others explore changing infrastructure and see what kinds of demands might emerge. Some are quick and dirty, others detailed and nuanced. And the models themselves have a temporal life that is affected by intellectual jealousy, competition, and collaboration that varies between modelling communities. And through this complex field, we repeatedly met modellers who are well aware of the limitations of their models and the concept of modelling, who realise how contradictory different modelling approaches are and how counter-intuitive is the notion of integrating different modelling types. Integration is on the political agenda, nationally and internationally, and significant amounts of research money are currently supporting work on integration activities. Primary among these activities is the modelling of integrated systems, while the modelling of integration itself relies on social scientists observing, interviewing, and to a limited degree participating in the activities of modelling itself.

Discussion

Each of the case studies highlights how the promise of energy infrastructures can be derailed when they meet the reality of local detail in all its messiness, its distance from the ideal, and the demand for small steps rather than grand leaps. In the German case, citizens were promised that everyone would be included in the energy transition, a promise that became an assumption that everyone was also in favour of proposals. Such an assumption, based on a well-meaning or morally positive-sounding basis, made it increasingly difficult to question policies if 'we' are assumed already to have agreed to them. In other words, there are many ways by which infrastructural change becomes politicised. Temporal conflicts themselves have a tendency to generate disillusion and disaffection. As Vike notes, as the twentieth century drew to a close, many liberal nations in the global North began to lose hold of their post-WWII welfare states (2013). These states grew out of a period of radical change built on hope, gathering momentum through the promise of an ideal future to come. After two generations, citizens began to ask why it had still not arrived, shifting from lending their agency to the creation of service to a demand to receive services. Vike characterises this as a shift from utopian time to contemporary time. When it comes to imagining energy infrastructures, our case studies suggest that energy temporalities are, in fact, struggling to negotiate between a utopian, decarbonised, endlessly renewable non-extractive future, and a contemporary struggle with quality of life, global and local inequality, and between global and local catastrophes[25].

---

25 Utopia, we remember, was always imaginary, a future with authoritarian tendencies rejected by its contemporaries (More 1516).

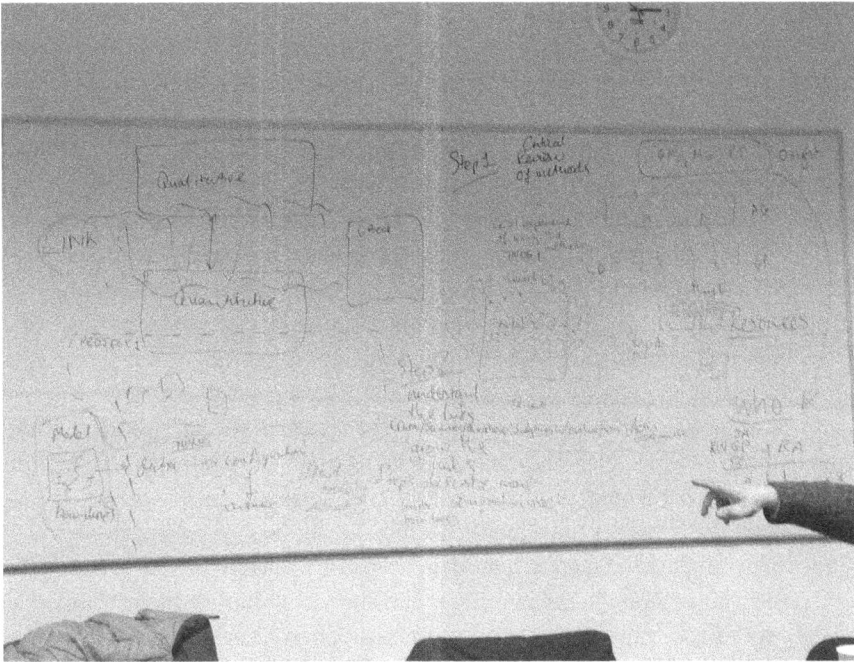

Fig. 3.15: Envisioning models. Credit: Abram

Perhaps surprisingly, the world of energy modellers is not particularly marked by idealism, at least not in relation to the extent to which models mimic the world. On the contrary, there is a realism attached to system modelling that acknowledges its partiality, its limitations, and its utility. Modelling is a kind of essential passage point for new energy infrastructures and technologies, and its idealism lies in the modellers' belief that it serves a useful function, one that is worthwhile and valuable. Questions of the ethical basis of modelling choices, or the moral justification for model design, were discussed with integrity among the modellers Abram and Silvast worked with, but remained external to the models themselves.

Next steps

Together these case studies illustrate the centrality of governmental technologies in normalising particular future-visions and delegitimising others. They highlight the infrastructural violence involved in transition practices, and the 'economies of expectation' that are cultivated in the process. They also illustrate the insights to be gained from careful ethnographic investigation of situations and practices that too often are summarily judged and simplistically represented in popular media as well as in governing settings, exacerbating conflict rather than moving toward resolution.

What does all this imply for future anthropologies of energy, and for anthropologies of futures? There is no dispute that 'saving the planet' is implicit in shared future visions, yet the futures of energy generation remain contested, politically, technologically, and in terms of who bears the costs. Anthropological research is well placed to unfold the arguments, practices, and consequences of imagined, trialled, and implemented energy futures. It draws on research into fossil fuels and renewable energy, but pays heed to the morally charged concerns around climate change.

The ethnographic approach suggests that we observe the aphorism 'more haste, less speed'. That is, it is valuable to take the time to investigate a situation in depth before rushing to conclusions or judgements. In this context, that means working to hinder the depoliticisation of disputes over energy futures. Anthropology offers the means to hold debates open, to maintain spaces for different positions, without moralising, stigmatising or excluding them. Ethnographers, working within and between those positioned differently around energy conflicts, can aim to understand the dynamics of conflict, whether between the priorities suggested by different temporal horizons, or between different moral or cosmological positions. Anthropological analysis helps to slow the arguments around urgency, relativising calls to 'save the planet', to highlight smaller but no less important concerns.

This is not to suggest that the role is either comfortable or necessarily achievable. But rather than cast anthropology in the role of saviour, we can open up the other positions that have value, such as moving towards pragmatism, and 'good enough' rather than ideal outcomes. Anthropological insights might help to foster partial solutions, and smaller steps. Although less heroic, we argue that there can be much to commend a modest pragmatism, perhaps rather like the modellers described in our case study, who know their work is not a perfect reflection of reality, yet has value and purpose that is worth pursuing. In contrast, then, to authors like the late David MacKay (2008), who argued that small steps were insignificant, we argue on the contrary that the grand scale does not exist, other than in the small actions that contribute towards the impression of grandiosity. Instead, by slowing down and paying attention to the particular, we get a better understanding of the whole, and a clearer view of the way that small steps lead to greater goals.

Conclusion

In summary, among the energy disputes discussed in this chapter, location and scale are often central in making disputes: between global vision and local effect, for example, or between the morally prioritised greater goals of saving the planet for humanity versus the sometimes trivialised concerns and anxieties of those whose expected futures and everyday practices will be usurped. In these scalar questions, temporalities are made and remade, and futures unimagined and reimagined, demoralised and remoralised. Renewable energy is tied into moral discourses about the future of the planet – on the global stage, yet locally it may threaten delicate ecosystems, vulnerable societies, or local economies, and often unwittingly reproducing extractivist and colonial structures.

These differently scaled energy futures are attached to different ethical stances and moral values, and related to different social positions. Is it more immoral to defy an assumed official consensus that future energies must be renewable, than to deprive industrial workers or fishing communities of their livelihood? Such incompatible arguments turn energy futures into a mire of different temporal horizons and dynamics, some linked to nostalgic tradition, others to care for future generations, or presenting a conflict over whether safety is tied more to continuity or urgent change. What draws together the different case studies presented in this chapter is the recognition that they rely on processes that can be understood to be technologies of governance that mediate processes of change, attempting to resolve or neutralise conflicts of interest. At the same time, these mechanisms try to align the competing scales of energy futures; with large-scale climate arguments depoliticising conflicting energy possibilities. Throughout the chapter, we have asked how official or normative future models, scenarios, and narratives are created, mediated, perceived, and contested by those who engage directly with them one way or another, whether among groups considered expert or not. Energy futures are revealed to occupy different presents, and to generate different forms of time. As transition theorists have argued for some time, there is no single transition and certainly not from one static state to another. On the contrary, there are multiple transitions, multiple temporalities, and a rich diversity of future imaginaries associated with energy and its generation.

References

Abram, S. (2016). Jokkmokk: rapacity and resistance in Sápmi. In G. Huggan and L Jensen (eds.), *Postcolonial perspectives on the European High North*. London: Palgrave Macmillan. 67–92.

Abram, S., and Weszkalnys, G. (2013). *Elusive promises: planning in the contemporary world*. Oxford: Berghahn.

Acosta, A. (2013). Extractivism and neo-extractivism: two sides of the same curse. In M. Lang, L. Fernando, and N. Buxton (eds.), *Beyond development: alternative visions from Latin America*. Quito: Permanent Working Group on Alternatives to Development/Fundación Rosa Luxembourg. 62.

Anand, N., Gupta, A., and Appel, H. (2018). *The promise of infrastructure*. Durham, NC: Duke University Press.

Appel, H. (2012). Walls and white elephants: oil extraction, responsibility, and infrastructural violence in Equatorial Guinea. *Ethnography*, 13(4), 439–465.

Argenti, N., and Knight, D. (2015). Sun, wind and the rebirth of extractive economies: renewable energy investment and metanarratives of crisis in Greece. *JRAI*, 21(4), 781–802.

Boyer, D. (2014). Energopower: an introduction. *Anthropological quarterly*, 87 (2), 309–334.

Breglia, L. (2013). *Living with oil: promises, peaks and declines on Mexico's Gulf Coast*. Austin: University of Texas Press.

Canino, M. V., and Strønen, I. Å. (2015). Oil and environmental injustice in Venezuela: an ethnographic study of Punta Cardón. In J. A. McNeish, A. Borchgrevink, and O. Logan (eds.), *Contested powers: the politics of energy and development in Latin America*. London: Zed Books. 116–140.

Cepek, M. (2018). *Life in oil: Cofán survival in the petroleum fields of Amazonia*. Austin: University of Texas Press.

Chien, K. H. (2020). Pacing for renewable energy development: the developmental state in Taiwan's offshore wind power. *Annals of the American Association of Geographers*, 110(3), 793–807.

Collier, S., Cross, J., Redfield, P., and Street, A. (2017). Preface: little development devices/humanitarian goods. *Limn 9*.

Cross, J. (2012). History, science and society. *Low carbon energy for development network symposium*. Loughborough University.

Cross, J. (2017). Solar basics. *Limn 9*.

Dale, B. (2016). Governing resources, governing mentalities. Petroleum and the Norwegian integrated ecosystem-based management plan for the Barents and Lofoten seas in 2011. *The extractive industries and society*, 3, 9–16.

Dunlap, A. (2018a). Counterinsurgency for wind energy: the Bíi Hioxo wind park in Juchitán, Mexico. *The journal of peasant studies*, 4(3), 630–652.

Dunlap, A. (2018b). The 'solution' is now the 'problem': wind energy, colonisation and the 'genocide–ecocide nexus' in the Isthmus of Tehuantepec, Oaxaca. *The international journal of human rights*, 22(4), 550–573.

Escobar, A. (1995). *Encountering development. The making and unmaking of the Third World*. Princeton, NJ: Princeton University Press.

Everts, J., and Müller, K. (2020). Riskscapes, politics of scaling and climate change: towards the post-carbon society? *Cambridge journal of regions, economy and Society*, 13(2), 253–266.

Feigenbaum, A., Frenzel, F., and McCurdy, P. (2013). *Protest camps*. London: Zed Books.

Ferguson, J. (1994). *The anti-politics machine: 'development', depoliticization and bureaucratic power in Lesotho*. Minneapolis: University of Minnesota Press.

Folch, C. (2013). Surveillance and state violence in Stroessner's Paraguay: Itaipu hydroelectric dam, archive of terror. *American anthropologist*, 116(1), 44–57.

Forde, E. (2020). *Living off-grid in Wales: eco-villages in policy and practice*. Cardiff: University of Wales Press.

Franquesa, J. (2018). *Power struggles: dignity, value and the renewable energy frontier in Spain*. Bloomington, IN: Indiana University Press.

Gandhi, M. K., and Jack, H. A. (1994). *The Gandhi reader. A sourcebook of his life and writings*. New York: Grove Press.

Gao, A. M. Z., Huang, C. H., Lin, J. C., and Su, W. N. (2021). Review of recent offshore wind power strategy in Taiwan: onshore wind power comparison. *Energy strategy reviews*, 38, 100747.

Gilberthorpe, E., and Rajak, D. (2017). The anthropology of extraction: critical perspectives on the resource curse, *The journal of development studies*, 53(2), 186–204. doi: 10.1080/00220388.2016.1160064.

Goodman, J., Chakrabarty, D., Connor, L., Gosh, D., Kohli, K., Marshall, J., Menon, M., Morton, T., Müller, K., Pearse, R., and Rosewarne, S. (2020). *Beyond the coal rush. A turning point for global energy and climate policy?* Cambridge: Cambridge University Press.

Gore, C. (2017). *Electricity in Africa: the politics of transformation in Uganda*. (African issues, 45.) Woodbridge: James Currey.

Hann, C. (2002). *Postsocialism: ideas, ideologies and practices in Eurasia*. London: Routledge.

Howe, C. (2019). *Ecologics: wind and power in the Anthropocene*. Durham, NC: Duke University Press.

Howe, C., Boyer, D., and Barrera Pineda, E. (2015). Los márgenes del Estado al viento: autonomía y desarrollo de energías renovables en el sur de México. *The journal of Latin American and Caribbean anthropology*, 20(2), 285–307.

Jasanoff, S., and Kim, S-H. (2013). Sociotechnical imaginaries and national energy policies. *Science as Culture* 22(2), 189–196.

Jensen, K. B. (2019). Can the Mekong speak? On hydropower, models and 'thing-power'. In T. Loloum, S. Abram, and N. Ortar (eds.), *Ethnographies of power: a political anthropology of energy*. NY: Berghahn Books. 121–137.

Loloum, T., Abram, S., and Ortar, N. (eds.) (2021). *Ethnographies of power: a political anthropology of energy*. New York: Berghahn Books.

Lord, A., and Rest, M. (2021). Nepal's water, the people's investment? Hydropolitical volumes and speculative refrains. In T. Loloum, S. Abram, and N. Ortar (eds.), *Ethnographies of power: a political anthropology of energy*. New York: Berghahn Books. 81–108.

Loulou, R., Goldstein, G., Kanudia, A., Lettila, A., and Remme, U. (2016). *Documentation for the TIMESModel*. IEA-ETSAP.

MacKay, D. (2008). *Sustainable energy without the hot air*. Cambridge: UIT.

Mau, S. (2019). *Lütten Klein – Leben in der ostdeutschen Transformationsgesellschaft.* Berlin: Suhrkamp.

Mitchell, T. (2009). *Carbon democracy, economy and society,* 38(3), 399–432.

More, T. (1516). *Utopia.* [Louvain]: Arte Theodorici Martini.

Müller, K. (2017). Heimat, Kohle, Umwelt. Argumente im Protest und der Befürwortung von Braunkohleförderung in der Lausitz. *Zeitschrift für Umweltpolitik und Umweltrecht,* 2017 (3), 213–228.

Müller, K. (2018). Mining, time and protest: dealing with waiting in German coal-mine planning. *The extractive industries and society,* 6(1), 1–7.

Müller, K., and Morton, T. (2016). Lusatia and the coal conundrum: the lived experience of the German *Energiewende. Energy policy,* 99, 277–287.

Murdoch, J., and Abram, S. (2002). *Rationalities of planning.* Aldershot: Ashgate.

Nader, L. (1972). Up the anthropologist: perspective gained from studying up. In Dell Hymes (ed.), *Reinventing anthropology.* New York: Pantheon Books. 284–344.

Nash, J. (1979). *We eat the mines and the mines eat us: dependency and exploitation in Bolivian tin mines.* New York: Columbia University Press.

National Grid (2017). *Future energy scenarios.* http://fes.nationalgrid.com/media/1253/final-fes-2017-updated-interactive-pdf-44-amended.pdf.

National Grid (2018). *Future energy scenarios (system operator).* http://fes.nationalgrid.com/media/1363/fes-interactive-version-final.pdf.

Nuijten, M. (2003). *Power, community and the state: the political anthropology of organisation in Mexico.* London: Pluto Press.

Oreskes, N., and Conway, E. M. (2010). *Merchants of doubt: how a handful of scientists obscured the truth on issues from tobacco smoke to global warming.* London: Bloomsbury.

Pandian, A. (2019). *A possible anthropology. Methods for uneasy times.* Durham, NC: Duke University Press.

Parker, W. (2014). Values and uncertainties in climate prediction, revisited. *Studies in history and philosophy of science,* 46, 24–30.

Parker, W., and Winsberg, E. (2018). Values and evidence: how models make a difference. *European journal of philosophy of science,* 8, 125–142.

Peterson, M., and Zwart, S. D. (2014). Introduction: values and norms in modeling. *Studies in history and philosophy of science,* 46, 1–2.

Rees, T. (2018). *After ethnos.* Durham, NC: Duke University Press.

Rudner, R. (1953). The scientist qua scientist makes value judgments. *Philosophy of science,* 20(1), 1–6.

Scheer, H. (1994). *A solar manifesto.* London: Earthscan.

Scheer, H. (2004). *The solar economy: renewable energy for a sustainable global future.* London and Sterling, VA: Earthscan.

Schlosberg, D. (2004). Reconceiving environmental justice: global movements and political theories. *Environmental Politics* 13(3), 517–540.

Schumacher, E. F. (1973). *Small is beautiful: a study of economics as if people mattered.* London: Blond and Briggs.

Scott, J. (1998). *Seeing like a state: how certain schemes to improve the human condition have failed.* New Haven, CT: Yale University Press.

Shever, I. (2012). *Resources for reform: oil and neoliberalism in Argentina*. Stanford, CA: Stanford University Press.

Smith, J. M. (2019). Boom to bust, ashes to (coal) dust: the contested ethics of energy exchanges in a declining US coal market. *JRAI (N.S.)*, 5(S1), 91–107.

Stevenson, E. G. J., and Kamski, B. (2021). Ethiopia's 'blue oil'? hydropower, irr(igation and development in the Omo–Turkana basin. In E. C. Gabbert, F. Gebresenbet, J. G. Galaty, and G. Schlee (eds.), *Lands of the future: anthropological perspectives on pastoralism, land deals and tropes of modernity in Eastern Africa*. (Integration and conflict studies, 23.) Oxford: Berghahn Books. 292–308.

Stewart, J. (2016). Visual culture studies and cultural sociology: extractive seeing. In *The Sage handbook of cultural sociology*. London: Sage. 322–334.

Strathern, M. (1996). Cutting the network. *JRAI*, 2(3), 517–535.

Vannini, P., and Taggart, J. (2014). *Off the grid: re-assembling domestic life*. London: Routledge.

Vike, H. (2013). Utopian time and contemporary time: temporal dimensions of planning and reform in the Norwegian Welfare State. In S. Abram and G. Weszkalnys (eds.), *Elusive promises*. Oxford: Berghahn Books. 25–55.

Warren, C. R., Lumsden, C., Richard, S., and Birnie, V. (2005). 'Green on green': public perceptions of wind power in Scotland and Ireland. *Journal of environmental planning and management*, 48(6), 853–875.

Weszkalnys, G. (2011). Cursed resources, or articulations of economic theory in the Gulf of Guinea. *Economy and society*, 40(3), 345–372. doi: 10.1080/03085147.2011.580177.

Weszkalnys, G. (2014). Anticipating oil: the temporal politics of a disaster yet to come. *The sociological review*, 62, 211–235.

Weszkalnys, G. (2016). A doubtful hope: resource affect in a future oil economy. *Journal of the Royal Anthropological Institute (N.S.)*, 127–146

Yang, C.-Y, Szerszynski, B., and Wynne, B. (2018). The making of power shortage: the sociotechnical imaginary of nationalist high modernism and its pragmatic rationality in electricity planning in Taiwan. *East Asian science, technology and society: an international journal*, 12, 277–308.

Zonabend, F. (1984). *The enduring memory. Time and history in a French village*. Manchester, Manchester University Press.

Nathalie Ortar, A. R. E. Taylor, Julia Velkova, Patrick Brodie,
Alix Johnson, Clément Marquet, Andrea Pollio, and Liza Cirolia

4 Powering 'smart' futures: data centres and the energy politics of digitalisation

Digital energy futures

From large-scale projects to compute climate data, socio-technical experiments with 'smart' energy grids and 'smart' cities, to the attempts of traditional energy companies to decarbonise through alternative fuels, post-carbon futures are imagined as digital and ethereal by corporate capitalism, governmental, and supragovernmental organisations. These imaginaries hinge on an old but still powerful idea that sees digital technologies and information flows as immaterial, and thus carrying the promise to enact 'post-industrial' societies that are not driven by the extraction and consumption of fossil fuels. Contemporary enactments of this idea flourish in public, industrial, and engineering discourse through imaginaries of 'sustainable' planetary futures moulded with the help of AI, predictive analytics, algorithmic-driven automated decision-making, and practices of data mining, all coordinated and produced in the cloud.

A vibrant and still emergent scholarship at the nexus of the anthropology of data and critical studies of media infrastructures has sought to go beyond these dominant narratives and to understand what shapes them. This scholarship has insisted on the crucial importance of adopting a situated perspective on cloud computing and the digital industries and infrastructures that support it, and have recognised the materialities and geologies of digital infrastructures and technologies (Hogan 2015, Dourish 2017, Brodie 2020a, Taffel 2021, Velkova 2021). It has explored the (geo)politics of their physical emplacement in particular geographies – through attentiveness to submarine fibre-optic cables, data centres, the mines of rare materials that power AI and cryptocurrencies (Starosielski 2015a, Lally, Kay, and Thatcher 2019, Crawford 2021, Johnson 2019). It has also drawn attention to what is discarded and left behind (Gabrys 2013, Liboiron and Lepawsky 2022, Brodie and Velkova 2021). This expansion of data, and the infrastructures, ecologies, and labours it requires, means we need to set a research agenda assessing present and future impact.

Within this emergent field of research, data centres – one of the most visible manifestations of cloud infrastructure – have provided a valuable entry point for exploring and engaging with new questions of the environmental impact of an increasingly digital world (see also Ch. 2). Inside data centres, the growing vol-

https://doi.org/10.1515/9783110745641-005

umes of data produced by the Internet-connected devices that we interact with on a daily basis are stored, processed and transferred. These buildings remotely deliver the apps, software, and computing resources that digital societies increasingly rely upon. Imagined as the 'engines of the Internet' (Alger 2013), data centres allow switching from one network to another and provide the computing and storage power that is essential to the smart digital futures being peddled by corporate strategists and government policymakers.

This chapter engages with these visions by questioning these smart and sustainable futures through a glaring blind spot: the data centres that underpin these futures are themselves energy-intensive enterprises. These buildings consume an incredible amount of energy to power and cool the computing equipment they contain. This can place burdens on local energy suppliers, leading to energy shortages for the neighbouring communities, as shown by Clément Marquet, Andrea Pollio, and Liza Cirolia in this chapter, thus reshaping local energy futures.

This chapter also intervenes by engaging with the unforeseen futures created by the extraction and use of natural resources and the reworking of local geographies entailed by the construction and running of data centres. Amidst growing public and political awareness about the environmental costs of data centres, data centre providers are engaging in efforts to 'green' their infrastructure. As such, considering data centres as an empirical object foregrounds the ways in which digitalisation, in its current form, both exacerbates problems of energy provision and environmental pollution, and fundamentally refigures and transforms local energy cultures (Strauss, Rupp, and Love 2013) and infrastructures, creating new relational entanglements between energy and data as argued here by Patrick Brodie and Alix Johnson.

In doing so we draw on ethnographic research to demonstrate the importance of the past in defining the shape of emerging systems, and identifying the pressure points at which they are disrupted, which will invariably shape the future. To engage with this ethnographically we explore relations of energy and data through the energy problems, politics, and promises of the data centres that underpin 'smart' futures through four cases. The cases investigate countries across the global North and South, in heavily urbanised as well as in less densely populated areas, addressing the presumed opacity of data centres by gaining access to these built spaces as well as around them, while also conducting research with municipal officials, energy operators and their neighbouring communities, among others.

Data centre energy: illusory anticipation

Data centres require electricity to power the hundreds (often thousands) of servers that they contain, as well as the air-conditioning systems that are used to keep the servers cool. These buildings have been estimated to consume 200 terawatt hours (TWh) of electricity each year – more than the national energy consumption of some countries (Jones 2018: 163–164). In doing so, they can cost their operators around $4 million per annum in electricity alone (Taylor 2018) and significantly raise the electricity consumption of whole countries (Danish Energy Agency 2018). Servers are far more energy efficient than they were a decade ago. However, due to their increasing deployment, they often consume the same, if not more, energy. This is a good illustration of Jevons's (1865) paradox, which, as Sy Taffel (2021: 11) succinctly explains, occurs when 'gains from increasingly efficient use of a resource fail to result in reduced consumption of that resource.' As of 2018, data centres were estimated to account for 1% of global electricity demand and to contribute roughly 0.3% to overall carbon emissions (Jones 2018: 163–164).

While data centres form the foundations of the digital future, they are often still powered by carbon fuel. For example, the average European renewable uptake was greater than that of fossil fuels in 2020, at 37% and 38% respectively (Eurostat 2022), although this varies considerably from country to country. Moreover, renewables are no panacea (see also Ch. 3), as they are dependent upon atmospheric movement and meteorological conditions. Concerns thus abound in the data centre industry as to the stability of the electricity provided by renewable power sources.

Other resources are also needed to power data centre equipment. For instance, the air conditioners require large amounts of water to cool the servers. Water has been identified as a key actant within data centre ecologies (Hogan 2015). During water shortages or droughts – which will increase in some parts of the world as climate change accelerates – the servers would quickly overheat, causing service disruptions and failures (Hogan 2015, 2018, Gilmore and Troutman 2020). A situated perspective on the cloud and data centres reveals energy systems as increasingly integrated with digital technologies for measuring, tracking, and distributing water, air and electricity, as well as forecasting consumption and generation. At the same time, as some data centres make the switch to renewable energy sources, they increasingly become entangled with elements of the environment that are often perceived in the industry as less predictable, such as wind (Bresnihan and Brodie 2021), thus revealing unforecasted futures.

Preparedness arises as a guiding logic of the data centre industry, providing data centres operators with a framework for anticipating and managing threatening futures (Taylor 2021a, 2022a). With any downtime potentially resulting in significant loss of revenue, data centre providers strive to ensure that their services will continue to be delivered under any circumstance by anticipating a range of speculative disaster scenarios – from earthquakes and hurricanes to unprecedented spikes in demand. This anticipation of future disaster often results in excessive amounts of energy consumption, with data centres leaving their equipment idling on standby in case there should be a sudden surge in demand (Holt and Vonderau 2015: 83, Taylor 2021b: 78). Preparedness leads to overprovisioning, with excess capacity remaining a standard operating condition (Taylor 2021b). Tung-Hui Hu (2017: 83) has thus suggested that data centre energy practices are driven by visions of future need translated into the present: 'The equation that converts an imagined crisis in the future into present capacity is why the cloud wastes so much energy.'

The politics of placing data centres

Alongside scaling up carbon-powered data centres, Big Tech (predominantly the 'Silicon 6' – Google, Apple, Amazon, Microsoft, Facebook, and Netflix) have sought to green their infrastructure in order to maintain and secure their market relevance and capital positions during the large-scale global transition to renewable energy. One recurring strategy has involved placing data centres in 'naturally cool' regions abundant in hydro- and wind power such as the Nordic countries, Ireland, the Netherlands, the US Pacific Northwest, and Québec. In these regions data centre providers can take advantage of the cold or windy climate to help keep their servers cool, reducing their cooling requirements and costs (Vonderau 2018). This has led Asta Vonderau (2019a) to observe that, within the regional investment strategies of Nordic municipalities, cold air is increasingly being reconceptualised as a natural resource that can be used to attract data centre development, with the hope this will lead to regional development.

The corporate rhetoric surrounding these developments has followed the same template and mobilised similar language, focusing on strategies of 'climate compensation' and 'carbon offsetting,' which have taken two forms. One has concentrated on building data centres in Europe powered by 'local, clean' energy, while the other on building wind and solar power plants owned by Big Tech that would match the capacity of the fossil-fuelled data centres in the US, turning Big Tech into important players in the transforming field of renewable energy provision and becoming energy providers. Research has demonstrated how these developments are generative of new frontiers of extraction that entangle renewable sources of energy built by local, public investments with the supply chains of global data capital (Bresnihan and Brodie 2021), thus simultaneously reproducing and extending earlier paths of colonial-driven capital accumulation and bypassing national communities. The contours of these emerging energy and data geographies and their implications for futures are only beginning to be mapped by social scientists, requiring further analysis for which this chapter seeks to lay the groundwork.

After the global financial crisis of 2008, European politicians and corporate lobbies helped frame data centres as new digital industries that could revive local economies (Brodie 2021). In response, national governments started to compete with each other, by zoning disused or abandoned industrial landscapes in urban and rural peripheries, or areas considered to be 'wastelands' into 'data-centre-ready' land, and reducing the cost of electricity needed to power data centres. This regional competition has given rise to a market landscape every bit as speculative as the forms of property development that drove the financial crisis in the first place. The ever-changing market dynamics and corporate strategies for computing geographies has left behind a trail of abandoned data centres or 'cloud ruins' (Brodie and Velkova 2020).

Ethnographic engagements and encounters at sites of data centre development and industrial exchange have been a key aspect of many studies of the data centre industry, whether among local communities affected by data centres (Burrell 2020, Brodie 2020b, Maguire, Watts, and Winthereik 2021, Vonderau 2019b) or industrial settings (Burrell 2020, Brodie 2021, Johnson 2019, Velkova 2019). They shed light on the social conflicts related to the extent to which local politics and common goods have animated local identity struggles bound to place and locality (Brodie 2020b, Burrell 2020, Mayer 2021, Vonderau 2018). Big Tech have extended social tensions and divides, reanimating historical conflicts embedded in the historical legacy of internal colonial dynamics and rural underdevelopment. As it quickly becomes apparent that data centres do not significantly improve regional welfare, local communities have tried to find value in other aspects of data centre operations. A range of initiatives are underway that aim to direct the waste heat generated by servers into district heating systems (Velkova 2016). This enables data centres to reduce their waste footprint while remaining firmly grounded in the carbon regime that powers their servers. As Julia Velkova (2021: 665) argues, these energy initiatives, 'simultaneously serve computing machines, the platform economy and old energy monopolies, while not necessarily breaking apart from the carbon regime.' If datafication is going to play a central role in energy futures, then it is increasingly important that data centre providers are brought into discussions with energy providers, municipal leaders, and urban planners. Data centres will need to feature far more prominently than they currently do in strategies and designs for more sustainable energy futures.

The cases

We present cases of contested data centre energy futures from around the world, including France (Clément Marquet), Iceland (Alix Johnson), South Africa (Andrea Pollio and Liza Cirolia), and Ireland (Patrick Brodie). These cases demonstrate the value of ethnography as a mode of accessing the social, cultural, political, and environmental side effects surrounding data centres, and exploring how their presence reshapes the future. Marquet's long-term ethnography of the Plaine Commune territory, a deprived region of Paris, offers an illustration of energy tensions at the local level, tracing frictions between electricity providers, local residents, and the data centres that have taken root in this urban region. These tensions are also present in Johnson's case about the data centre industry in Iceland, despite the country's abundance of renewable energy which attracts companies wishing to green their consumption. Johnson draws our attention to the political construction of Iceland as an energy island defined by its geographic isolation and its abundance of renewable energy. Pollio and Cirolia highlight the complexity of running one of Cape Town's first co-location data centres in an energy-deprived country, revealing its paradoxical dependency on past infrastructures and its invisibility. Path dependency shapes data centre energy presents and futures. This is also clear in Brodie's case, which unravels the tangled web of past and present circumstances on and around a peat landslide in rural Ireland caused by the construction of an access road developed to provide renewable wind energy for data centres in Dublin.

The cases highlight ethnographically the tensions between data centres, their energy consumption, and their uneven geographical distribution. They reveal the inconsistencies in public policies, the weakness of legislative powers to counteract the social and environmental costs of data centres and the new geographical and social inequalities they create. In doing so the cases emphasise the importance of ethnographic analysis as a tool for investigating how Big Tech companies participate in the production of futures framed by digital technologies which increase energy consumption and exacerbate energy inequalities. Thus, engaging with dominant visions of smart energy futures critically, the cloud and data centres become situated through the political-economic and colonial relations that are constituted by the hunt for cheap energy and resources to power them (see also Ch. 5), and the pre-existing power asymmetries and forms of exclusion these relations entail. As such, the cases shed light both on how data centres reconfigure futures and how ethnography can bring unheard voices and alternative futures to local and national governments and other stakeholders.

Contested data centres and energy futures in Paris' Plaine Commune

Clément Marquet

It is 15 January 2015: around 80 people are gathering in rue Rateau in La Courneuve, in the north of Paris, a rather poor suburban city part of the Plaine Commune urban community, known for high unemployment rates, immigration, and energy insecurity. As the crowd stops, Jade Lindgaard, a journalist and ecological activist, takes the microphone to say:

> I invite you to turn towards this metallic grey building and to look at it carefully. It is interesting to compare this building, its aesthetics, its size, to the other buildings of this street, namely houses, which date from some 20 years ago, of one or two floors, facing this metallic parallelepiped which dominates them and breaks the aesthetics of the street (figure 1). You can't see it from the outside, but it's a data centre. A data centre of the company Interxion, but no sign indicates it. From our point of view, this invisibility is part of the problem of data centres.

This fieldwork snapshot is an excerpt from a 'Toxic Tour' organised by a local ecologist group. Toxic Tours are a demonstration technique (Barry 2001) used by environmental justice activists to attract attention to the correlation between polluting infrastructure and inequalities. The Toxic Tours organised in Plaine Commune aimed at showing the entanglement between climate and social and environmental inequalities, and that the territory was gridded by polluting and noisy infrastructure, motor roads, airports, and data centres which did not benefit the local citizens and caused environmental damage.

Unknown to the citizens and most of the elected officials, data centres have been silently mushrooming for years, thus considered as 'invisible' by the Toxic Tour activists. Between 1999 and 2015, fifteen data centres thrived in a rather small area, leading politicians to promote their territory as 'the French capital for the data centre industry' in 2009 and journalists to qualify the area as the 'golden triangle of cloud computing in France' (Dupont-Calbot 2013), a reference to the shape of the former industrial area, bordered by a railway in the west, the

Fig. 4.1: Interxion Data Centre Par7 facing the houses of the rue Rateau. Photo by author.

Parisian ring road in the south, and the Saint-Denis canal in the north-east, crossing Aubervilliers and Saint-Denis.

Between 2011 and 2015, the proliferation of these infrastructures started to raise concerns about their energy implications, their economic contribution, and their urban integration: local energy providers, environmental agencies, financial services, and city dwellers sent alerts to the municipality. As a result, in La Courneuve the operator has been sued and two Toxic Tours organised to uncover the issue to a wider audience.

Drawing on the sociology of public problems (Dewey 1927, Cefaï and Terzi 2012) and on urban infrastructure studies (Larkin 2013, Graham and McFarlane 2015), this case explores the micropolitics of trouble (Emerson and Messinger 1977) regarding data centres installations, while issues related to climate change and the energy transition became increasingly important in urban policies and for residents. The case unravels how energy infrastructure, energy precarity, and space allocation became central in the debates and rendered data centres politically visible in Plaine Commune, and how environmental critics failed to produce political changes regarding data centres. I have conducted twenty interviews between 2015 and 2018 with elected officials, public bodies, neighbours, data centres consultants, and architects. Since then, I have also conducted observations through the regular attendance of data centres professional fairs in France, and analysed the grey literature (reports, environmental impact assessments).

From energy abundance to the threat of virtual energy shortage

Only in 2009 did local elected officials start to pay attention to the data centres settled in Plaine Commune and turned those infrastructures into an asset for local economic growth. Jean-Yves Vannier, a freshly elected official in charge of Urban Planning, Economic Development, and Sustainable Development, met several experts in data centres to discuss the development in the city of Aubervilliers. As data centres generate important revenue for local taxes and occupy warehouses taken up by activities providing little revenues like wholesale businesses, he envisaged that data centres could become an asset for urban planning and a tool for economic gentrification. To carry out this project, he reached an agreement with Data4, a French operator of neutral co-location data centres, and planned a 15,500 m² data centre.

In 2011, the energy provider Enedis warned the Plaine Commune elected officials that energy was lacking in the territory, and suspended all new data centre installation until the building of a new power substation. This was mostly due to the practice of overbooking electricity, an economic strategy that takes advantage of Enedis's reservation system for economic purposes: 'I realised that the data centre operators were waging an incredible war on each other by reserving electrical power to block the competitor's development,' Jean-Yves Vannier told me. Unable to exclude economic competitors by purchasing land as too many brown fields were still available, data centre operators have exploited the electricity reservation system to gain an economic advantage in Plaine Commune's golden triangle.

The story might have ended with the provider's alert and the suspension of data centre installations. However, Jean-Yves Vannier, eager to have a new data centre in Aubervilliers, made the issue public. 'In order not to have the technical contingencies as a political compass', he asked the administration of Plaine Commune and the energy supplier for the construction of a new energy substation, infrastructure planned and built in a longer time-frame than data centres. To force the decision, he called journalists and contributed to the publication of a press release entitled 'Energy shortage threatens Plaine Commune's data centres' (Le Parisien 2011). While the article made Enedis's lack of provision responsible for slowing local economic development, it started to raise the administration and city dwellers' concerns about data centre power consumption.

Tax reform and energy transition

While Jean-Yves Vannier was urging the administration to invest in a new substation, data centre growth became a matter of concern and in 2010, Plaine Commune's Urban Ecology Department was asked to implement the Territorial Climate Energy Plan (TCEP), a program with quantified objectives to act locally on climate issues.

Defined by European guidelines, these plans aim to reduce greenhouse gas emissions and energy consumption by 20%, and to reach a threshold of at least 20% renewable energy in final consumption by 2020 at the intermunicipal level. In an interview, the territorial civil servant in charge of the TCEP mentioned the contradictions in which he finds himself as he is asked to encourage already energy-poor residents to reduce their energy consumption, while elected representatives allow energy-consuming infrastructure to settle without any debate. And, as data centres were bringing revenues through local taxes, they were almost unchallengeable.

In 2012, a report analysing the economic impact of the 2009 national reform of the business tax revealed that the economic contribution of data centres was drastically lower than expected as, between 2009 and 2011, data centres tax revenues had dropped from €12.6 m to €1.5 m, an 88% decrease (Service des études financières 2012). The note questioned the role of data centres in Plaine Commune's fiscal resources, as 'data centres require a lot of space and energy'. The Urban Ecology Department of Plaine Commune used this alert to investigate data centres.

In 2013, territorial officers and a local environmental agency produced the first French expert report analysing the technical, urban, and ecological impacts of data centres (Leicher 2013). This report had three objectives. First, to provide a technical understanding of data centres, as most of the elected officials and territorial officers had little knowledge about this activity. Second, to create a debate between the elected officials of Plaine Commune in order to define a common political ground regarding the territorial climate commitments and future data centres. Third, to propose instruments to support such a policy position and to regulate the future installation of data centres: the authors suggested constraining the location of data centres to reuse waste heat and suggested new environmental indicators in order to assess the collective and environmental contributions of companies and make data centres comparable to other projects.

However, the presentation of this report to the Plaine Commune Board of Directors raised strong differences between the elected officials, as those in favour of data centres argued that they were essential to economic development (and therefore should not be regulated). Since the disagreements were too great, the absence of consensus favoured the pro-data centre actors.

From local protests to climate mobilisation

The tensions surrounding the data centres did not remain within the administration and these infrastructures began to be contested on the street in a fight opposing the rue Rateau dwellers to Interxion's data centre Par7, a co-location neutral data centre (Fig. 2). The Interxion data centre was built in two stages, 2011–2012 and 2013–2014, in order to avoid an administrative procedure requiring environmental authorisation due to the hazardous nature of the data centre as the building stores nearly 580,000 litres of fuel to power generators in the event of a power outage, a feature requiring a longer administrative procedure, including a public inquiry.

During the summer of 2013, an inquiry took place while the first half of the building was already commissioned. Residents felt cheated as they learned that the massive, ugly, and noisy building facing their home was also a potential fire hazard or could even explode. In response, they launched a petition which obtained 424 signatures, and took Interxion to court with an environmental law firm.

As their struggle began to be publicised, the climate activists who were organising the Toxic Tours in the Seine-Saint-Denis department joined them. Data centres were reported as consuming a lot of energy and land without creating many jobs in a territory known for low incomes, energy precarity, and high unemployment. Moreover, the Seine-Saint-Denis has a long history of the struggle against environmental damage caused by motorways and airports. Data centres were presented as polluting and undesirable infrastructure contributing to climate change.

In October 2015, the case against Interxion Par7 took place. At first the residents won as the court invalidated the procedure and suspended the ongoing authorisation. This decision implied that Interxion's clients would have to withdraw their servers within a few days because of a procedural error. However, the local administration softened the decision two days later: Interxion could pursue its activities, but had to produce another procedure in due form.

Fig. 4.2: In the background, Par7 the data centre of Interxion, overhanging the houses of the rue Rateau. Photo by author.

Epilogue

In 2021, the story of data centres in Plaine is still ongoing. Two hundred metres away from the rue Rateau, Interxion has started to build a new data centre campus in La Courneuve (Fig. 3). This campus has been celebrated by the operator in the media as one of the largest sites in Europe. In a local press article, the journalist quoted an Interxion executive, asserting: 'We are drying up the electrical capacity of the sector' (Debruyne 2020) thus notifying its competitors that no other data centre could settle in the area, due to a lack of power in the local grid. As the biggest data centre in the area, the campus secured 130 MW of electrical capacity. Answering to local critiques, Interxion explained its willingness to 'offer heat to whoever wanted it' (Debruyne 2020) – without saying who would pay to build a heat network locally – invest in training for unemployed people to provide jobs for the data centre industry, and build an 8,000 m² park which would be retrofitted in the city.

Fig. 4.3: First building (of four) of Interxion's future data centre campus, construction site in La Courneuve, April 2021. Photo by author.

Discussion

Through the dynamics at play since the late 1990s in Paris's northern suburbs, this case highlights how data centres are at the centre of contested territorial and energy futures as well as the relationships between data centres and energy temporalities.

First, it sheds light on how, amidst political and public blindness, data centres contribute to infrastructure concentration and impact the energy grid and supply. Data centres are an infrastructural reactivation of 'dormant' energy supply left vacant by older industries (Pickren 2017). The authorities hope to reap some benefits from their growth by promoting them as carriers of a new modernity, a revival of the Industrial Revolution that once brought wealth and employment (Vonderau 2019b, Maguire and Winthereik 2019). In the name of this new modernity, marketed as a revival of old industrial prosperity (be it true or not), these cloud infrastructures of the future are welcomed regardless of their needs in energy infrastructure, which are taken for granted (Lally, Kay, and Thatcher

2019). Yet, as data centres appeal for investments in energy infrastructure and foster the development of telecom infrastructure (fibre optic cables), they reinforce infrastructural concentration and path dependency mechanisms (see Brodie's case). But it is only when energy starts to be a matter of concern for various actors (future data centre operators, elected officials, and the national energy provider) that the relationship between data centres and energy becomes political and a subject of public and political inquiry.

Second, conflicting modes of anticipation appear between the energy provider Enedis and the data centre industry. While Enedis uses an anticipation instrument, the energy booking system, to manage local energy demands and forecast investments in power substations on a long-term and national level, data centres overprovision electricity to answer clients' demand (security, scalability, availability) and to secure future developments against economic competition. Energy issues regarding data centre growth and network planning are spreading across Europe: Amsterdam recently declared a moratorium on data centre installations (Judge 2020), and regulating data centres is a growing concern in Frankfurt (Judge 2021) and Dublin (Swinhoe 2021). What is at stake here is the production of new modes of anticipation (instruments, regulations) that can address the practices and politics of data centre energy and land management.

Third, the case highlights the frictions carried by the unrestrained spread of data centres and their uneven consequences. The multiple failures of the actors to problematise the link between data centres and energy, and their inability to produce a public debate about other territorial futures, stress the difficulty of protesting against this 'infrastructure of modernity.' By highlighting the potential danger to neighbouring communities, its inability to meet municipal climate commitments, and its contribution to climate injustices, the critics framed data centre development as an undesirable future at the local and global level and produced alternative narratives of territorial energy and climate futures. They proposed instruments (from the ecological indicators to the Toxic Tours) to make those alternative sociotechnical imaginaries exist. Nevertheless, each attempt to contest, expose, or inflect the consequences of data centres energy practices for the territorial future has been countered by arguments linking economic development and technological determinism. If, thanks to the mediatisation of local protests and the multiplication of expert reports, operators now must deal with their public image and accept trade-offs in local negotiations, the future energy challenges are still secondary for the operators and the collectivity while data centres secure their own energy future.

The energy island

Alix Johnson

The first time I felt the interplay of hot and cold air inside a data centre, it immediately called to mind – or, more accurately, to body – another new experience I was having in Iceland, where my research on these digital infrastructures was based: the cross-currents reminded me of the delicious indulgence of cranking up a radiator while opening a window to the winter winds. Icelanders, I had already learned, did this often, as a way of enjoying some fresh air in a season spent overwhelmingly and unavoidably inside. Raised as I was with an attitude of Midwestern frugality, and in an era of energy conservation campaigns, I was shocked by the idea of letting heat out of a warm house. But Icelanders, with their crisp and cosy homes quickly convinced me – in Iceland, unlike any other country I had lived in, neither the cost, nor the carbon footprint of their heating (provided through geothermally warmed water), was a concern. So I had started, tentatively and then with abandon, cultivating this atmosphere inside my own apartment, savouring the feeling of falling asleep amidst a light breeze. It was this feeling that returned to me standing in the data hall, reaching around to feel the heat seep out behind the server rack, while cool air rose steadily from the floor beneath my feet. But unlike the exchange of air inside my apartment, in the data hall heat and cold were precisely controlled.

As Mateo, my guide that day, explained: 'the principle of data centre design is separating and containing hot and cold air.' Because computation generates heat as a by-product, and too much heat stresses servers and other hardware, cooling is a central concern for data centre operators – and a major expenditure. Meticulously separating the hot air produced by servers from the cold air used to cool them down makes striking such a balance more efficient and cost-effective. So, at the data centre in Southern Iceland where Mateo (an American veteran of the data centre industry) worked as Director of Technical Services, servers were organised into a common industry configuration of 'hot' and 'cold aisles': servers are stacked on racks, which are arranged in back-to-back pairs, and every other row is sealed within its own little glass room. Cold air, filtered up through the floor, is blown into the front of each rack, while the hot air emitted from the back of the servers is trapped in the aisle behind them, then channelled up and out of the room. This delicately engineered environment, Mateo told me, keeps servers running reliably within a thermal window of 68 and 71 degrees Fahrenheit (20 – 21.7 degrees Celsius).

So while my own moment of embodied recognition in the data centre conjured a condition of energy abundance (the ability to run a heater against wide open windows), Mateo's explanation of the data hall's organisation drew my attention to practices of energy containment, instead. As it turns out, and as I wish to explore in this short piece, both dynamics are essential to the project of drawing data centres to Iceland in the first place. Both constitute Iceland as a kind of energy island, a place known for energy production that is unlike, and crucially *unlinked to*, anyplace else.

Over the past decade, scholars and industry observers have noted a northward migration in the data storage industry (Holt and Vonderau 2015, Vonderau 2019a). This is because the work of cooling servers is made easier if the environment surrounding the data centre is already cold: where other sites have to pay for air conditioning, in Iceland Mateo likes to say, 'we just open the windows' (although, of course, as we have already seen, the reality is somewhat more technically involved). In addition to its cold, Iceland's appeal as a data centre site is enhanced by its abundance of renewable energy: its national grid is comprised of 73% hydroelectric and 27% geothermal power, and produces 55,000 kWh per person annually, as compared to the European Union average of 6,000 kWh (Government of Iceland n.d.). This energy abundance is reflected in promotional materials that aim to attract the industry to Iceland: commissioned reports, brochures, and websites all cite statistics surrounded by sumptuous natural imagery, rushing rivers and steamy rifts that signal the promise of power. Energy abundance is substantiated in the twelve-year contracts the national power company is able to offer data centres, as the pricing of renewables is more predictable than that of fossil fuels.

But it is not only Iceland's capacity for energy production that matters to data centre developers (and the clients they seek to attract): the isolation of Iceland's grid counts for something too. In a public talk Mateo gave in 2016, at an industry conference called World Hosting Days, he explicitly contrasted Iceland with mainland Europe: The European power grid, he explained, 'is interconnected by seventy per cent. I'll translate that for you: a power outage or disturbance in Macedonia or Portugal can take out parts of Germany – it has happened, and it will continue to happen.' In Iceland, by contrast, Mateo said: 'You're an island.' This means that Iceland's energy is 'non-dependent on any imports, non-dependent on any other grid.' In framing the island as wholly separate and self-sustaining, Mateo emphasised that it was insulated from risk – Iceland's servers would stay online, whatever happened in Macedonia or Portugal.

At the same time, others have pointed to the ways that interconnection compromises, while Iceland's isolation keeps its power 'pure'. For example, one Icelandic cloud services provider coined the moniker 'Truly Green' to distinguish itself from the competition: while other companies claimed to be 'greening' on the basis of carbon credits that offset emissions, or mixed grids with a growing percentage of renewable power, Iceland's almost fully renewable – and completely independent – grid meant that data stored and processed by this company was 'truly green' in a way that others could not claim (Høvsgaard Nielsen 2014). It is, then, not only Iceland's abundance of energy that makes it appealing to the data centre industry; it is also, perhaps equally, the containment of that energy – the construction of Iceland as an energy island.

By 'energy island' I mean that Iceland is an island associated with energy production, but also that Iceland is an energy producer associated with isolation. Unlike the complexly interconnected grids of Europe (which Mateo warns against in his talk), Icelandic energy is produced and consumed here. It is, in fact, this place boundedness that makes 'Icelandic energy' legible as a concept at all: Icelandic energy is that derived from landforms on the island, and put to use within its national borders, which coincide neatly with the extent of those landforms. Like another set of energy islands, the Orkneys, which Laura Watts (2019) traverses in *Energy at the End of the World*, energy here emerges at the volatile nexus of environment and innovative engineering; through the relationship between a people and a place experienced as an earthly 'edge'. But unlike in the Orkney Islands, no cable connects Icelandic energy to any mainland; no contracts make Icelandic energy accessible outside. Instead, the specific benefits, and situated nature of that energy are* what draws data centres (and others) to Iceland: they actually need to be on the island to partake in its promises, to access the abundance available here. Doing so, they avoid the risks, and sidestep the PR pitfalls, of more distributed sources of energy. Iceland's islandness would seem to offer the data storage industry an exception to business as usual; an energy elsewhere or escape.

Islandness, however, is not just a self-evident geographical state of being; it is also a freighted cultural imaginary. The fact of Iceland's small size, relative remoteness, and surroundedness on all sides by the North Atlantic Ocean have long been mobilised and made meaningful, from both inside and out. In antiquity, Iceland was associated with Ultima Thule, the conceptual endpoint of the knowable world; later, visitors to the island would emphasise its exotic difference and distance from the European mainland (Ísleifsson 2011). In Iceland's independence movement, nationalists invoked their persistence amidst the harsh conditions of island living as evidence of both a distinctive Icelandic identity, and the capacity of Icelanders to govern themselves (Oslund 2011). More recently, Iceland has been productively associated with the trope of the island laboratory, for example when the genetics company deCODE proposed to construct a national database of all Icelanders' medical, genealogical, and genetic information. As CEO Kári Stefánsson put it, Iceland's small and geographically isolated population made the island 'perhaps the ideal genetic laboratory' (Greenhough 2006).

But another trope of island living was activated after the construction of that genealogical database, when a team of software engineers won a design competition with 'Íslendinga-App,' an application that let users bump their phones together to determine if they were too closely related to date. 'The Icelandic nation is not inbred,' Kári felt compelled to tell *USA Today* following a spate of international media coverage of the 'anti-incest app': 'This app is interesting. It makes the data much more available. But the idea that it will be used by young people to make sure they don't marry their cousins is of much more interest to the press than a reflection of reality' (Associated Press 2013).

All this is to say that Icelandic islandness is not only a property of its land mass and location; islandness has also served as a framing device for claims and characterisations; it has sometimes functioned as a liability, and sometimes as a strategic resource. The 'lure of the island' (Perón 2004) does important work towards determining what kind of place Iceland is, and what kinds of purposes it can be said to serve. Particular notions of islandness get reified in these manoeuvres, from familiar notions of remoteness to the naturalisation of Icelandic nationhood. Today, key players in the data storage industry partake in this iterative inscription process, picking up on island imaginaries as they emphasise Iceland's geographical isolation. As Iceland is marketed as energy 'non-dependent' or as uniquely and 'truly green,' the island is cast as a space of exception where the regular rules do not apply. Reinforcing ideas of Icelandic distance and disconnection in order to carve out a new economic niche, the lines they draw around the island feel intuitive, as they trace in the grooves of lines drawn before. In this formulation, Iceland emerges not so much as a model of the renewable energy transition, an example to which other places could aspire; instead, it appears as outside, an otherwise, an escape hatch for an industry incentivised to cut both its costs and its carbon emissions, fast.

And yet. As scholars (and inhabitants) of islands have long argued, 'islands and continents are but names we give to different parts of one interconnected world' (Gillis 2004: 3). Data centres energy futures, likewise, are not so easily cordoned off or parcelled out. While the Icelandic national grid may be contained on the island, Icelandic energy both does work, and relies upon relationships with others, outside its shores.

These relationships might be said to have started in the 1960s with the arrival of the Alusuisse (which later became Rio Tinto Alcan) aluminium smelter at Straumsvík on Iceland's south-west coast. An infamously energy-intensive industry, smelters seek out inexpensive power contracts and move their operations accordingly; they were drawn to Iceland at the time for its hydropower, which drastically reduced energy costs as compared to oil. While early negotiations triggered controversy in Iceland over the growing influence of multinational corporations (Iceland, after all, had only gained full status as an independent republic in 1944), the aluminium smelter was approved and erected in conjunction with a hydropower plant at Búrfell. This dual construction project guaranteed the Alusuisse smelter a reliable long-term source of inexpensive electricity, as it doubled Iceland's energy production capacity. At the same time, it set off a trajectory of twinned development – an aluminium plant for a smelter, a smelter for an aluminium plant – that has continued since. For example, in 2006 the Kárahnjúkur Hydropower Plant was built in conjunction with the Fjarðaál aluminium smelter (owned by Alcoa) in Reyðarfjörður on Iceland's east coast. More recently, geothermal drilling in the Hellisheiði area has also accelerated to meet aluminium smelters' needs (Maguire 2020). Each of these projects has been controversial in its own way – the Kárahnjúkur dam aroused protests for its environmental impacts (Magnason 2006) and the acceleration of geothermal drilling in Hellisheiði has produced a profusion of anthropogenic earthquakes in the region. But these linked projects of expanding infrastructure and expanding industry have produced Icelandic energy on an unprecedented scale; they have made Iceland the world's largest electricity producer per capita.

'Icelandic energy', then, has long been international. Of course, the energy itself, to borrow Watts's term, 'Icelandic electrons' (2018), is derived from Icelandic land forms and accessed primarily through Icelandic labour, innovation, and expertise. But the creation of 'Icelandic energy' as a branded commodity, and the development of the national energy grid into an appealing industrial resource, has been negotiated in relation to multinational companies, aluminium smelters specifically. It is through these relations that Iceland has become known as a kind of 'energy island' at all. From one angle (and one viably held viewpoint in Iceland), this development has been an economic success story: in 2018, aluminium smelting accounted for over 17% of Iceland's total exports (Iceland Chamber of Commerce 2019); it has been vital in diversifying an economy once tethered almost entirely to the sea. But from another angle, the outsized influence of multinationals that sceptics warned about in the 1960s has come to pass: heavy industry now consumes 77% of electricity in Iceland (Iceland Chamber of Commerce 2019), while benefiting from labour reforms, tax incentives, and environmental exemptions (Skúlason and Hayter 1998).

If in the second half of the twentieth century, Icelandic energy was primarily marketed as being inexpensive (an influential 1995 pamphlet produced to attract power-intensive industry was straightforwardly titled 'Lowest Energy Prices'), today Icelandic energy is known for being green. It is its almost entirely 100% renewable constitution that appeals to companies pressured to address their carbon footprints, such as the data centres increasingly approaching Iceland as a place where their electricity consumption can be both easily accommodated and conveniently contained. Environmental scholars and activists have long drawn attention to the ecological costs of even 'green' energy, the real harms done by damming and drilling, and the sacrifices these practices often entail. But here I wish to emphasise the work done, and the strategic opacities enacted, by the construction of Iceland as an energy island defined by its abundance and isolation.

Imagining and describing Iceland in this way positions the island as an escape: here, energy-intensive industry is shielded from the risks of interconnection, as well as exposure to public critique. Iceland appears an irresistible outside to norms and trade-offs that seem unavoidable elsewhere. But energy futures take their traction from energy histories, and the concept and function of Iceland as an energy island have been constructed through (often compromising) international relationships. As data centres step into a legacy largely made by the aluminium industry, they raise some of the same questions about foreign corporate interests and the application of Icelandic energy (Magnússon 2015). Today, data centres consume more energy than all the households on the island put together (and recall those open windows and radiators running on high) (Ingólfsson 2019). Based on anticipated industry expansion, power shortages are predicted in Iceland in the coming years (Stefánsdóttir 2019). And so, while the expansion of the data-storage industry has not yet commissioned its own new power plant – its own extension of Icelandic energy infrastructure in its interests – such a project has now been placed on the table, a plan for developing a hydropower plant in Iceland's West Fjords that would likely serve data centres ever-increasing needs (Guðmundsson 2019).

Iceland, then, is not an energy island 'naturally' suited to the data centre industry. As we have seen, the energy abundance on the island has been historically, and is continually being, engineered in the interests of others; meanwhile, the infrastructures that channel this energy are not 'isolated' but instead developed in intimate relation to international needs. Like the cross-currents of hot and cold air that hit me inside the data centre itself, Icelandic energy may appear a natural profusion, but is in fact carefully directed by design. Rather than a matter of 'just opening the windows,' data centres are practically sustained in Iceland by a dense network of interconnections. There is no escape then, no meaningful outside to the exorbitant and growing power demands of digital data. Offshoring data centre operations may make these less visible, but does not do away with the collective responsibility of managing their many costs. Much as we may long for an energy island, our energy futures are indelibly linked.

Engines of the Internet: South Africa's energy-data nexus

Andrea Pollio and Liza Cirolia

On a crisp July morning, we – three urban scholars and an energy expert – were consuming miniature cupcakes and coffee sitting in a windowless boardroom inside one of Cape Town's first co-location data centres. After several rounds of biometric security, we had been generously met by the head of marketing, accompanied by a small team of bubbly engineers. The centre had become, over recent years, Africa's largest neutral Internet Exchange (IX). Yet the reasons for our desire to visit the facility went beyond a mere curiosity with South Africa's recent Internet history. With local elections around the corner, two issues kept appearing in Cape Town's political debate: incessant load-shedding – that is, rolling electricity blackouts across the city and country[1] – and the development of Amazon's new African headquarters on a contested brownfield site just outside the city bowl, an area known as Two Rivers Urban Park (or TRUP for short).

While elections have since passed, load-shedding and debates over TRUP have not. And, although the two issues might seem unrelated, they are in fact two facets of the inextricable nexus of energy and data in Cape Town, a city which has played and continues to play a unique role in the development of the global cloud in Africa and beyond. As far as energy is concerned, our field trip was born out of a keen desire to understand how notoriously energy-hungry data centres function in a country where the generation and the distribution of electric power are in a seemingly endless crisis. Despite this, South Africa houses Africa's lion's share of data hosting capacity, with two-thirds of large-scale facilities on the continent (Pollio and Cirolla 2022a). Early in 2020, for example, Amazon announced that the city would be the site for the first African node of their global cloud services AWS, thanks to servers in three different facilities. Although these data centres remained unnamed, it was very likely that we were indeed visiting one of them. But there was more to Amazon's story: in Cape Town, a decade and a half earlier, a small team of Amazon software engineers developed EC2, the software backbone of AWS (Pollio and Cirolla 2022b). Facilities like the one we toured, therefore, attest at once to the recent, uniquely local, history of the cloud and to its energy futures.

---

1 In South Africa, load-shedding is the result of separate development policies during apartheid, which delivered an unequal infrastructuring of the country, and of decades of mismanagement of the country's national energy utility in the democratic era.

Our tour had begun with a lengthy PowerPoint presentation. As our hosts spoke, artists' impressions and aerial images of new facilities being built all over the country alternated on the screen. But unlike those renderings of new, large-scale, purpose-built premises, older data centres are often housed in very unremarkable buildings: concrete structures that previously functioned as telephone exchanges, former warehouses wedged between residential properties, or underground garages of commercial buildings. In our case, our trip had taken us to an office complex in Rondebosch, a leafy suburb perched on the slopes of Cape Town's famous Table Mountain. Surrounded by tall plane trees, by the 1950s the office building had replaced one of the colonial farmsteads which once dotted the area, and whose old Dutch and English names still identify streets and properties. Today, the refurbished complex hosts a mix of IT companies, asset management firms, a couple of law firms, and, in its underground belly, Africa's densest web of network switches.

To reach the data centre, we walked past the hallway of the building, with its glossy marble floors and whitewashed walls, and descended to a level below the ground. Compared to the opulent corporate feel of the levels above, the data centre interiors featured a mix of high-tech and very functional, no-nonsense aesthetics. While waiting for our hosts, a stream of couriers would come in and out, carrying boxes of various sizes and disappearing into the corridors. It felt like being at the entrance of a bustling underground anthill. 'In a way, we are the landlords of the Internet,' we were told by our host. But while the metaphor of the landlord evoked ideas of ownership, our hosts meant the exact opposite. 'As a neutral facility, we just rent out the space and make sure that the lights are on. We have no idea about what happens inside the server cabinets.' As an example, one of the engineers anecdotally mentioned a content distributor, one of the companies that are licensed by media firms such as Netflix to make their content available outside their native jurisdictions. He told us that they had come with their own cabinet and that if somebody tampered with it or touched it, the servers inside would wipe themselves and self-destruct. The idea seemed quite extreme, but we guessed it was also a cautionary tale so that we should not touch anything during our later visit. We all naïvely pictured one of us poking a little yellow cable protruding from a cabinet and shutting off Netflix for the whole continent.

And so our conversation went on for several hours. Our energy expert took the lead. An established consultant working on just urban transitions and deeply concerned with the fracturing of urban systems, she asked detailed questions about the procurement and supply of the large amounts of electricity necessary to power the whole machine. With often unpredictable load-shedding – resulting in blackouts which would last for spans of two hours several times a day – ensuring continuous power supply, despite uncertainty, was a full-time job for our data centre. It turned out that the data centre reported paying Eskom, the national power utility, an exorbitant monthly bill. While the details of the contract were not disclosed to us – for example whether or not it included a reserved quota of electricity – large commercial and industrial premises all over the country are in a constant, usually unsuccessful, negotiation with Eskom to be spared from load-shedding. And although we had not yet ventured into the server rooms, we all felt that we had entered a big engine made by a network of smaller machines, each made of an even smaller webwork of switches, processors, disk drives, and power units singularly innervated by a lattice of electric cables.

The impression of having walked into an engine became even more intense when we were eventually guided into one of the server rooms. A loud whirring noise surrounded us, an imperceptible vibration almost indistinguishable from the dense, lukewarm air and the slightly noxious smell that made some of us pleasantly light-headed and others increasingly anxious and claustrophobic. This low murmur is the result of a very simple phenomenon. Processing units are, after all, conductors of electricity. As electrons travel through the unit, they collide with other particles inside the medium, releasing energy in the form of heat. The more a processing unit heats up, the less it is capable of performing, and the more likely it is to lose data. Therefore, as described by Johnson in her case, each server is equipped with small fans that push warm air out of the server cabinets into the so-called hot aisles, from which this exhaust air is extracted, filtered, and cooled down before being recycled. Cool air is then fanned into the cold aisles, narrow corridors onto which the server cabinets open. The low humming noise is the compounded result of all the fans and the air-conditioning units that continuously keep each part of the 'engine' at its appropriate temperature. The containment of hot and cold air in these narrow aisles is the latest architectural technology in what has been a very fast evolution of how data centres have become better and better at reducing their energy consumption. And yet, as our data needs increase, so does the energy needed for hosting them an industry that already churns out 1% of global carbon emissions (Masanet, Shehabi, and Lei 2020). Africa's contribution may be a very small portion, but is also the fastest-growing destination for data hosting, and questions about how this industry's needs for power and water will be met remain to be answered.

Scholars of digital infrastructure have indeed variously shown how data depend on other resource-intensive grids: not just energy, but also water supply (Hogan 2015), and, in African cities, alternative networks of distribution (Guma 2019). The geography of data also follows older infrastructures, from railways (Burrington 2015) to telegraph lines (Starosielski 2015a, 2015b). Shannon Mattern has argued that there is a spatial path dependency between new and older media systems (Mattern 2017). For this reason, while global tech corporations are experimenting with relocating data facilities underwater or in remote Arctic sites, older facilities like the one we visited are here to stay. The dense web of connections that have been built, incrementally, as the centres have grown, would be incredibly difficult to move geographically. We realised this as we walked past the IX room, a nondescript door with a rounded porthole window. Our guide did not let us in, but explained that the matrix of connections inside was so inextricable that it would be impossible to physically move to a new location. So while purpose-built facilities can be assembled in more efficient ways (even underwater!), old switching networks may remain for the foreseeable future in the ageing buildings that first housed them. But there is more. Not only are these knotted networks here to stay, but what seems to be the thermodynamic relationship at their core is also here to stay.[2]

Understanding this relationship begins with taking seriously the sensory experience of having entered an engine. Engines are human artefacts that transform power into work. While the mechanical work of a traditional engine and computing work are obviously not the same, the notions of computing data and work are so entwined that the first time the word computer was ever used it referred to (female) workers (Light 1999). If data centres are industrial-size engines doing computing work, then perhaps they are not too dissimilar to modern factories, which emerged in the late eighteenth century as an architectural and logistical project to organise the relationship between two forms of work: human labour, and mechanical work produced by the steam engine (Daggett 2019). Data centres still have this relationship at their core – the optimisation of different forms of work and the energy required for them (Hu 2015: 89 – 90). Therefore data centres may become more efficient, but their energy futures are still couched in their pasts, in the social and physical thermodynamics of work and energy. In this sense, data centres like the one we visited are not just revealing of the fact that the internet is a very spatial matter, but also of the relationship between computing and energy, which too depend on the infrastructures that have built cities over decades. These fixities, therefore, at once depend on and inform what Louise Amoore has called 'cloud geographies' (2018).

In many African cities, these geographies inevitably intersect with the path dependencies of colonial spatial planning, which resulted in only partial urban

---

2 At least until quantum computing becomes the norm, although some speculate that superconductors (i.e. processing units that do not produce heat because they are kept in a cryogenic state) will never fully replace normal CPUs.

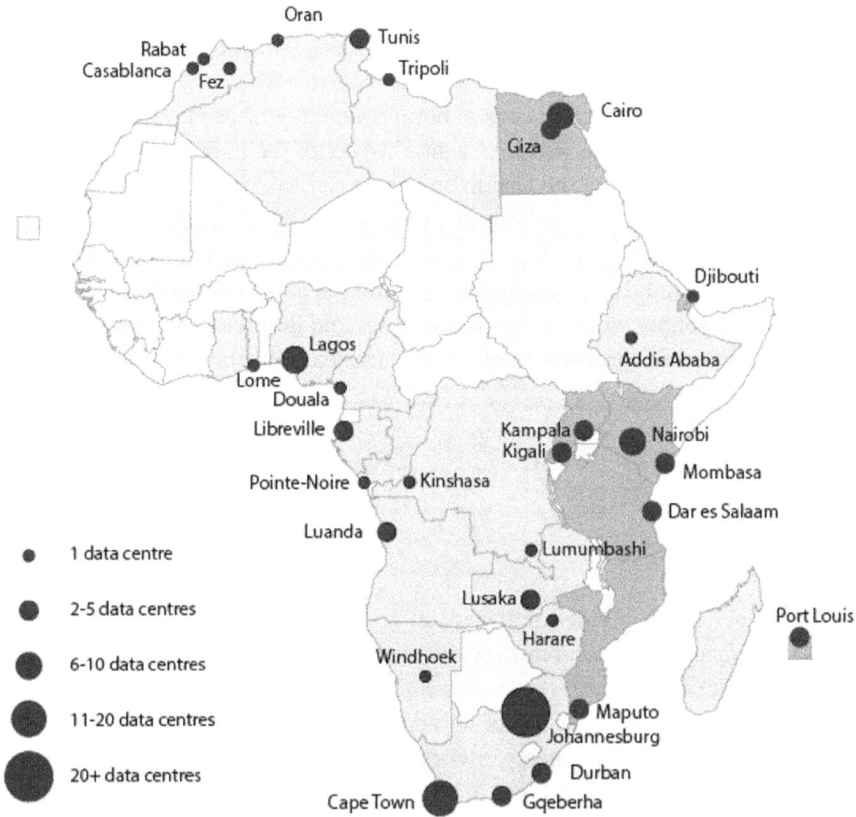

Fig. 4.4: The uneven (and) urban geography of co-location data centres in Africa as of August 2021. The booming data hosting industry in Africa, due to spatial legacies of colonial infrastructure, is an almost exclusively urban phenomenon. Map drawn by the author based on the Cloudscene database.

networks for grid infrastructure. In Cape Town specifically, the need to have centrally located data centres also circles back to Amazon's unique history in the city, where the software behind the first commercial cloud was developed and where thousands of software engineers keep being hired to experiment with making remote computing work more efficient. Similarly, the data centres that host Amazon's servers keep building energy-efficient resilience against load-shedding. The nexus of data and energy is what anchors the cloud to specific locations in Africa. As we were writing this sketch, Amazon announced the construction of a large solar plant to power its servers in a country and in a continent that still struggles to meet energy requirements for much more basic needs. The irony of these entanglements is that one doesn't need to travel a long distance from the data centre we visited to find large suburbs where the infrastructural promises (Von Schnitzler 2016) of the South African constitution are far from being a reality.

As we left the data centre, several hours after entering, we looked back at the old building between the withered winter leaves of the plane trees: a building we'd seen many times on the way to our university campus but never really noticed. We also passed the neighbouring care home, whose guests with advanced hearing aids, we had been told, kept complaining about the obnoxious and endless whirring sound that would haunt their unique auditory frequency. We could not help but wonder whether they'd ever been told that the sound was the murmur of the Internet, the whir of one of its engines converting energy into work.

Clouds and bogs

Patrick Brodie

In November 2020, a video surfaced on social media of a catastrophic peat landslide adjacent to a wind farm construction site in County Donegal in the Republic of Ireland (ROI) near the border with Northern Ireland (NI). The landslide was caused by the construction of an access road for the Meenbog Wind Farm being developed by Invis Energy, a Cork-based joint venture by Irish- and English-based stakeholders, who had in 2019 entered into a contract with Amazon Web Services (AWS) to provide energy for their data centres in Dublin, serving the US multinational's commitments to renewable energy. The access road, like the planned wind farm, was being built over a vast blanket bog landscape. In press materials, AWS are largely absent from discussion about the landslide, in spite of only months before being praised for their climate responsibility and investments with regards to the wind park (Quann 2019) – in fact, it appears as though their links to the landslide were scrubbed from some initial reports, and they have directed all media inquiries towards Invis Energy (Corr 2020).

This ethnographic case unravels the tangled web of circumstances surrounding this peat landslide and its aftermath through the lens of current and historical infrastructures and ecologies. To do so, I will present a series of empirical vignettes through which we can understand the entanglements of culture, land, territory, resources, people, and politics when faced with the externalities of data centres' energy demands and their administration along infrastructural networks colonised by private capital. Most of this research occurred during the 2020 (and 2021) Covid-19 pandemic, where many researchers have found that '"the field" withdraws further and further' from us (Vemuri 2022), and most data collection has happened via Twitter, frantic Googling, and poring over any policy, planning, and media documents available. Thus much of the research is reflexive/reflective, narrativised through examples that seem disconnected but work together via the intensities and entanglements of Ireland's histories and infrastructures. That said, I was lucky enough to briefly visit the site of the Meenbog Wind Farm construction site and landslide with a colleague on a wet, rainy day in June 2021 and speak informally with some people involved in protesting against it and other large-scale infrastructures being developed in the region, which afforded the opportunity to photograph and reflect on the place with my boots on the soggy ground and learn from the experiences of those who knew that something like this could happen.

Path dependence

Fig. 4.5: High capacity pylons crossing the Barnesmore Gap in Donegal, Ireland. Photo by author.

My colleague and I were stopped at an automated traffic signal for construction on the N15 at the Barnesmore Gap in Donegal. We were not sure what the roadworks were, but I was struck the day before, when I had passed through in the other direction, towards the west coast of Ireland from the north, that a trail of high capacity overhead electrical pylons followed alongside the road, which itself followed along a decommissioned railway track, most recently under the governance of the County Donegal Joint Railways Committee from 1906 to 1960, which incorporated the earlier West Donegal Railway connecting Stranorlar in the mountainous Finn Valley with the town of Donegal closer to the West Coast. The line had first been established in the late 1800s by the British colonial authorities, who oversaw the creation of an extensive railway network connecting the reaches of the largely rural, agrarian island of Ireland, making accessible its various resources, agricultural, and labour forces.

The Gap itself was much more ancient, its dramatic location between two peaks in the Bluestack Mountains partially carved by the retreat of glaciers cutting across the once-frozen landscape 13,000 years ago, creating what today appears as a natural passageway across the landscape. The blanket bog which carpets the rock formations underlying the Gap and its surrounding mountains is a direct ecological remnant of these retreating glaciers, as thousands of years of low-lying plant decomposition – most prominently the Sphagnum, or 'bog-builder', moss which creates a pillowy bed across the underlying, semi-aqueous peat reserves – created the complex soil formations that have long been harvested and used for household cooking and heating and industrial energy and agricultural purposes. Swathes of these blanket bogs are still beholden to the ancient Turbary Rights of local residents, or the right to use certain plots of land for turf-cutting, and veins of drainage, excavation, and stacks of dried peat made by small-scale turf cutters are visible for miles across many of these largely uninhabited landscapes, even in the shadow of 115 m high wind turbines and alongside sprawling evergreen plantations.

The existing energy, transport, and industrial infrastructure thus followed in a layered path dependence of prehistoric geology, pre-colonial settlements, colonial governance, and postcolonial development, wherein the conditions of the 'Anthropocene' and its modes of production and uses of resources entangle the atmosphere, land, infrastructure, and people. These existing routes of development are increasingly becoming important for Ireland's 'green' transition.

Wind farms, like data centres, are built along and to fit into these existing networks and pathways; often assembled along the border and/or on otherwise undeveloped bog landscapes, they must be assembled on land that is suitable (in the case of bogs, land seen as good for nothing else), but must also be at least marginally linked via existing infrastructure to transport enormous turbines for installation and electrical cables to distribute electricity to substations, storage, and the grid. Where these networks do not already exist (and often even where they do), they must be fashioned through enormously transformative and disruptive means by, for example, widening rural roads, creating new access roads through inaccessible landscapes, laying new cables, constructing new pylons, and building new substations. Thus the distributed networks in which wind farms take part, far from 'undergirding' the everyday life of locals in rural Ireland, have tended to disrupt or bypass them, whether overhead pylons transporting energy elsewhere, roads closed, or expanded only for large lorry transport, or fibre-optic cables underfoot bringing data to and from Dublin and the oceanic cable connections on the west and north coasts.

Deep peat

Peat is a carbon reserve; when dried and extracted it acts as a hot burning, dirty carbon fuel for individual or industrial activity. But while remaining in the ground it actively stores atmospheric carbon. Intact peat bogs in Ireland, as of 2005, stored 1,085 megatons of carbon, equivalent to 53% of Ireland's soil carbon stores in only 16% of the land area (Irish Peatland Conservation Council 2020). For this reason, bogs have come to the attention of many environmental scientists, especially those involved in carbon accounting, over the past few decades as a viable site of carbon sequestration. Left in a 'natural' state, peat bogs can be made valuable as 'sinks' for human pollution and be mathematically used to offset carbon emissions elsewhere. Once they are drained and cut, much of that carbon is already released into the atmosphere, from the ground into the air. When these are rewetted, they will eventually become active, and store carbon once again. For this reason, government grants are now being offered to farmers and landholders to leave their bogs alone, not developing them or using them for grazing or turf-cutting – whether in the interest of local or regulated conservation or the planetary science of sequestration.

The large-scale, high-tech infrastructures of renewables and Big Tech exist in stark contrast to the forms of underdevelopment that still characterise much of the country, especially the oft-disconnected border region, whether lack of broadband access or other connectivity. This is an energy story as much as an access story, as energy systems transform to accommodate more 'responsible' energy consumption. The non-fossil fuel central heating systems, for example, are unevenly available; – in 2017, four out of ten households in Ireland still used oil for heating, but 66.4% of households in the border region used this form of fuel while only 11% did in Dublin. Similar to oil, homes in border counties Leitrim (8.9%) and Donegal (11.1%) still rely heavily on peat for heating compared with only 5.4% countrywide (Central Statistics Office 2017). This high level of individual household carbon demonstrates the ongoing reliance on localised, small 'e' energy, as Larry Lohmann calls it (2016), while large-scale renewable energy remains unavailable for heating unless new technologies (e.g. for retrofitting) become affordable and widely available, for now pooled into a grid that can heat their computer's battery but not the air in their home.

We got back in the car and followed the pathway of the N15 to the Meenbog Wind Farm construction site, which was just on the other side of the construction on an access road with a small sign. We were hugging the border – only a few miles across the mountains, past the job site, was County Tyrone, in NI, and the site of the landslide. Access to the job site was restricted, unless you were a turf-cutter that needed access to the surrounding bogs to take advantage of turbary rights. Works were meant to be halted on the site while an investigation into the cause of the landslide was ongoing, but several cars passed through the gates as we wandered. Signs marked warnings against trespassers and environmental hazards in the area as we left the site: 'Deep peat.'

Fig. 4.6: Sign cautioning about 'deep peat' at the entrance to the Meenbog Wind Farm construction site in Donegal, Ireland. Photo by author.

'Like a blister'

The Meenbog Wind Farm has been under development since 2016, and has been a subject of great controversy, especially in Finn Valley, where the Finn Valley Wind Action Group has been lodging complaints and objections against other local wind farm developments since at least 2015. In 2019, while planning was still underway for Meenbog, AWS and then-Taoiseach (Prime Minister of ROI) Leo Varadkar announced publicly that the company would be buying 100% of the projected 91 MW of energy produced by the wind farm, another step in fulfilling their pledge to eventual 100% renewable energy use. In this arrangement, wind energy does not go directly to AWS' data centres, rather pooling in the grid with other energy, and AWS' energy contracts would simply ensure that some of that ends up with them, allowing them to put on paper that they are 'committed' to renewable energy while avoiding any direct responsibility for its development (either positively or negatively rendered). In his statement, however, Varadkar publicly celebrated the deal as a direct example of AWS' climate commitments and contributions to the Irish economy (Quann 2019). The project is now one of three wind farms covered by corporate power purchase agreements entered into by AWS in Ireland, joined by projects in Cork and Galway.

Before AWS entered the picture, the planning application for 49 wind turbines at Meenbog was initially denied after a series of environmental appeals, leading another wing of Invis' parent company to apply for a second, smaller-scale planning permission for 19 turbines. This second attempt was successful, in spite of earlier zoning rules forbidding wind farm development due to 'proven environmental sensitivity' (Corr 2020), demonstrating the often competing sustainability claims mobilised through the bogs. AWS only entered into the project once the red tape had mostly been cut and environmental impact assessments completed (however partially), to avoid public service levies, as the project would be entirely unsubsidised by public funds. Locals were 'baffled' to hear that AWS was involved in the project (Kiernan 2019), especially after so much controversy, and coming from the mouth of no less than the Taoiseach of Ireland, paving the way for approval before any had actually been reached. To their knowledge, AWS officials had not directly engaged in any public hearing or consultation with local communities, signifying that AWS' 'investment' in the wind farm was a contract with another corporation and not with local interests.

The Finn Valley group as well as the Gweebarra Conservation Group continued to raise serious concerns about the integrity of the peat bog on which the turbines would be built, and across which a network of access roads would have to be constructed. These roads across the bog are called 'floating roads', as they essentially sit on top of largely aqueous peat bog, which sits underneath most of the landscape on the surrounding mountains, whether planted with sitka spruce trees or left in a more natural bare state for grazing, turf-cutting, or conservation. Conventional wisdom from those most involved and knowledgeable about the landslide says that it was caused by the construction of the access road to one particular turbine upstream, from where it spilled into the Mourne Beg River. A digger excavating the ground had popped an underground reserve 'like a blister that ruptures' according to one hydrologist (qtd. McSweeney 2021), sending a layer of peat down the hill and into Tyrone in NI and its water supply and fisheries. We were told by residents that people fishing the water had pulled up catches with black peat clogging gills. The scale of the fish kill reportedly would be hard to estimate due to the scale of the peat pollution, which could have been obscuring death underneath. NI and ROI authorities both told residents not to worry about damaged water supplies.

Fig. 4.7: Aftermath of the peat landslide on a bogland site crossing Donegal, Ireland and Tyrone, Northern Ireland. Photo by author.

Walking on the hill to observe the fallout of the landslide, which was still extremely visible more than six months after the initial slippage, I was struck by just how remarkably unsuitable for construction this land must have been. Most of the bog essentially felt like walking on a waterbed, as you bounced and could see the ripples undulate for several metres away from your feet. Occasionally the bog became even squelchier, sucking your boot into it for stepping on a patch unsupported by plant roots. This was the muck that lay underneath, from 5 to 20 m deep, storing centuries of decayed plant material in its depths. To engineer this landscape for wind farm construction required excavation through the entire peat store, to get to the more stable rock formations underneath in order to plant concrete to support the turbines.

Conclusion

Inspired by Anne Pasek, who tells us that a Microsoft data centre complex in Virginia can be told through a story of 'concrete and coal ash' (2019), this particular data centre and energy story can be told perhaps through clouds and bogs, atmospheric and geological systems colonised by big tech capital. While appearing as 'externalities', they are actually constitutive of the systems themselves. The links in the data and energy supply chain traced above are becoming battlegrounds against Big Tech and its visions of a green future administered by their profit motives. Whether wind farms or pylons or battery storage, these links in the data and energy supply chain offer concrete sites of the struggle against a 'future' that we are told is already here.

Narrativising these connections is as important as mapping them, as it is often the connections between these sites that appear arcane or incomprehensible. But they are connected culturally, infrastructurally, and environmentally; people know these systems that cross through their communities and the landscapes around them, and how they connect with broader political and economic formations. An objector to a battery farm in border County Leitrim (ROI), proposed by Canadian multinational Brookfield Renewable Partners, could not only map out the watershed that would be affected in the case of storage failure, he could also point to where the energy lines coming in and out of the neighbouring substation went, which continued onwards from near Dromahair to a hydropower plant in Ballyshannon and then across the nearby border into NI. Border, energy, and nature stories coalesce within the ways that people understand the vast systems that govern and dictate their paths, while systemic forces like the state and corporations enact different forms of spatial transformation and expertise that frequently come into friction with existing ways of knowing and interacting with these places.

The reflections offered here may seem disconnected, or relatively scattered, as they unravel the entangled economic, cultural, political, and environmental factors involved. It may also seem that, in the peat bogs of rural Donegal and Tyrone, I have drifted far away from data centres, worlds away from the 'databelt' that surrounds Dublin. Even with the mere contrast of wind farms with the small-scale energy cultures of turf-cutting that coexist in these spaces, it seems more of an 'energy story' than a data centre story. In many ways, this is undeniably correct, as ultimately Amazon is simply creating demand for renewable energy that would, hopefully, be built anyway to substitute for declining fossil fuel reliance. Data centres may be an infrastructure peculiar to a particular kind of 'smart' and 'green' industrial formation emerging as the quintessential makeup of the 'Industrial Revolution 4.0', but they remain only one part of a vast system of energy and data infrastructures powering the data economy, private companies suctioning energy off public grids while proclaiming that they are shaping their constitution by developing renewable energy. In Ireland, as elsewhere, there is no separating the island's energy story from its data centre story anymore, nor from its industrial-scale green transformations. This eco-modern transformation will fundamentally reorganise Ireland's energy systems and its distribution of electricity across the island, something which will inevitably affect anywhere else that shares in the benefits and problems of the island's data and energy systems.

Even though Ireland is an island, its energy systems, along with their externalities, have political and material effects and influences far beyond its shores. Whereas in Iceland the abundant renewable energy is produced and contained within the island, Ireland's energy infrastructure remains carbon-intensive, in transition, and increasingly connected to wider energy systems as it is already connected to Great Britain by two existing undersea interconnectors. Island-wide renewable energy plans rest almost entirely on wind energy, and Brexit threatens the unity of approach that will be necessary for wholesale, island-wide transformation. In response, a planned interconnector between Ireland and France will plug Ireland directly into the European grid.

Next steps: energetic ethnographies of digital media infrastructures

Nathalie Ortar, A. R. E. Taylor, Julia Velkova, Patrick Brodie, Alix Johnson, Clément Marquet, Andrea Pollio, and Liza Cirolia

As anthropologists and social scientists engage in the analysis of the multifarious issues raised along the nexus of energy and data futures, they provide a much needed compass that helps reject instrumental, teleological views on digital futures. The issues that we have discussed in the introduction and the cases show how societal and planetary benefits from digitalisation and 'smart' futures are neither inherent to nor automatically produced by digital technologies and infrastructures. On the contrary, the emplacement and energopolitics of data centres (Boyer 2019, also see Ch. 3) in specific places reproduce in many respects well-trodden paths of industrial extraction. They create new issues for the communities at the centre of these developments by generating new social and identity conflicts and reactivating questions of modern infrastructural provision that have long been taken for granted in particular geographies. Taking inspiration from this scholarship, we propose five empirical interventions needed to expand these perspectives and move forward research on energy and data futures.

First, ethnography is uniquely positioned both to reveal the kind of invisibilities which are at stake and to mobilise the processes of making things visible which enables alternative futures to emerge. The data centre industry is commonly presented as a secret, opaque, and invisible industry (Blum 2012). Therefore we have underlined how those infrastructures tend to be both invisible (Graham 2016) and hypervisible, whether that is through the colourful pictures released by Google and Facebook (Holt and Vonderau 2015); through data centre marketing and promotional practices (Taylor 2021a); through negative press coverage about data centres' energy consumption; or through narratives of local and national business attractiveness. This work has demonstrated that multiple layers of in/visibility shape and structure the data centre industry and the futures that this industry underpins. In/visibility is a relational feature of infrastructures – what is invisible for some is often highly visible for others (Star and Ruhleder 1996) and what matters is how (in)visibility is mobilised, by whom, and why (Larkin 2013). To paraphrase Larkin, the politics and poetics of data centres emerge through the practices, troubles, and issues that make specific aspects of Internet infrastructures in/visible.

Second, we need to take local concerns and forms of agency, resistance, or debate seriously in discussions about the futures enabled by data centres and the in-

frastructures that help enact digitalisation. Policy and scholarly analysis have to pay closer attention to the silenced publics who are infrastructurally incorporated in new relations between the digital economy and the energy politics of climate change. The build-out of new energy required for these practices similarly follow fault lines of uneven industrialisation, meaning that energy generation required for the growth of data centres tends to offer minimal benefit to ordinary energy consumers like households and small businesses. Amplifying local voices and value conflicts illuminates ultimately what is at stake in the negotiation of power, both literal and metaphoric, as well as the conflicting values that enter into friction with the entanglement of data and energy.

Third, data centres' energy futures lie in past social and physical thermodynamics of work and energy while creating new patterns which are part of their ambiguity. Data centre operations are entangled and dependent on multiple infrastructures such as roads, abandoned industrial zones, heat and electricity provision plants, bogs and peat extraction farms, as well as the humans who keep all of these operational (Taylor 2019, 2022b). Taking these entanglements as a focal point of analysis lets us powerfully illuminate the socio-material and knowledge practises through which data centres reanimate questions of local energy transitions, environmental justice, infrastructural path dependencies, and manifold forms of colonialism. Data centres' energy futures invite us to foreground the spatial dynamics of path dependency patterns. Indeed, if 'older media networks have laid the foundation for our modern-day systems... the 'old' systems those we might regard as buried on the '"ower strata" – are also very much alive [...] in contemporary infrastructural systems' (Mattern 2017: xxviii). New patterns are also emerging in the making of energy and technology futures, as different infrastructures may converge at one moment, giving traction and form to these futures. Yet material convergence does not automatically lead to the integration of the different economies of value, time, and care that underpin each of these infrastructures separately. The divergence in the latter acts as a force of instability that ultimately may lead to the disentanglement of converging infrastructures, and the reshaping or demolition of the futures associated with them.

Fourth, the concept of 'energy gentrification' (Libertson, Velkova, and Palm 2021) can be mobilised as a critical intervention to illuminate how local actors and livelihoods are transformed through the anticipatory energy demand for data centres, affecting and limiting local access to electricity. Most data centres and the digital futures tied to them are built on preparedness (Taylor 2021b, 2022a). While Big Tech's solar and wind-production power plants produce more renewable energy, they do not expand the grid capacity. This raises concerns about the grid's capacity to meet the needs of public facilities such as hospitals. The energy gentrification perspective highlights the trade-offs between common goods versus private interests. It is thus fitting to ask, what other forms of gentrification related to the energy politics of data centres emerge?

Fifth, future research needs to reckon with the Euro/US centrism of much of the discourse and research produced on data centres so far and acknowledge the need to unpack the ongoing unevenness of global cloud infrastructure and its future use to produce alternative narratives. Future directions of this research will need to account for how the imagination of these uneven geographies has excluded the growing concentrations of data centres *within* the 'global south' and the new forms of inequalities between the global North and South. Data colonialism combines the predatory extractive practices of historical colonialism with the abstract quantification methods of computing (Couldry and Mejias 2019). It will provide the preconditions for a new stage of capitalism that as yet can barely be imagined, but for which the appropriation of human life through data will be central. Such developments implore us to understand capitalism's current dependence on this new type of appropriation that works at every point in space where people or things are attached to today's infrastructures of connection.

Conclusion

With the scale and reach of Big Tech's energy futures, a pressing question surfaces about the need to rethink the knowledge and approaches to data centres and their entanglement with energy infrastructures, by considering local and situated, human and non-human concerns, in order to empower local needs and communities. While in many ways the power enacted by Big Tech companies in energy systems is new, these processes of private power within public systems demonstrate the historical continuity of certain forms of value extraction and capitalist accumulation. The 'transition' we are witnessing from carbon-intensive modernity to 'green' industries and technologies in the post-industrial landscape has not, and will not, radically transform the foundational uneven circuits of capital that define current global systems of exchange. These systems will continue to enrol local communities and environments in different ways, along existing fault lines of access, exploitation, and abandonment. Acknowledging these facts requires a research agenda that will illuminate how the industrial infrastructures of the 'fourth Industrial Revolution' produce new extractive and energy-intensive industries and how this hampers futures in the name of technological progress, efficiency, and preparedness; a research agenda that ultimately gives voice to alternative futures.

By highlighting different ways of inverting (energy) infrastructure and data, the energy ethnographies of data centres we have discussed in this chapter illustrate how ethnography can be mobilised to disentangle the relations between materialities, flows, and data and energy capitalism.

Acknowledgements

We thank all the people who have participated in our research, since without their collaboration and commitment it would have been impossible to develop the cases discussed in this chapter. The research we discuss in this chapter was supported by the following funding organisations and research partnerships: Alix Johnson's research has been supported by funding from the National Science Foundation, the Wenner-Gren Foundation, the American-Scandinavian Foundation, the Leifur Eiríksson Foundation, and the American Council of Learned Societies. Patrick Brodie's research was carried out with the support of a Fonds de Recherche du Québec – Société et Culture Postdoctoral Fellowship (282690).

References

Alger, D. (2013). *The art of the data center*. Prentice Hall: Englewood Cliffs, NJ.
Amoore, L. (2018). Cloud geographies: computing, data, sovereignty. *Progress in human geography*, 42, 4 – 24.
Associated Press (2013, April 18). New app helps Icelanders avoid accidental incest. *USA Today*. https://eu.usatoday.com/story/tech/2013/04/18/new-app-helps-icelanders-avoid-accidental-incest/2093649/
Barry, A. (2001). *Political machines: governing a technological society*. London: Athlone Press.
Blum, A. (2012). *Tubes: a journey to the center of the Internet*. New York: Ecco.
Boyer, D. (2019). *Energopolitics: wind and power in the Anthropocene*. Durham, NC: Duke University Press.
Bresnihan, P., and Brodie, P. (2021). New extractive frontiers in Ireland and the Moebius strip of wind/data. *Environment and planning E: nature and space*, 4(4), 1645 – 1664. https://doi.org/10.1177/2514848620970121.
Brodie, P. (2020a). Climate extraction and supply chains of data. *Media, culture and society*, 42(7 – 8), 1095 – 1114. https://doi.org/10.1177/0163443720904601.
Brodie, P. (2020b). 'Stuck in mud in the fields of Athenry': Apple, territory, and popular politics. *Culture Machine*, 19. https://culturemachine.net/vol-19-media-populism/stuck-in-mud-in-the-fields-of-athenry-patrick-brodie/

Brodie, P. (2021). Hosting cultures: placing the global data center 'industry'. *Canadian journal of communication*, 46(2), 151–176. https://doi.org/10.22230/cjc. 2021v46n2a3773.

Brodie, P., and Velkova, J. (2020). Cloud ruins: Ericsson's Vaudreuil-Dorion data centre and infrastructural abandonment. *Information, Communication and Society*, 24(6), 869–885. https://doi.org/10.1080/1369118X.2021.1909099.

Burrell, J. (2020). On half-built assemblages: waiting for a data center in Prineville, Oregon. *Engaging science, technology, and society*, 6. https://doi.org/10.17351/ests2020.447.

Burrington, I. (2015). How railroad history shaped Internet history. *The Atlantic*, November 24.

Cefaï, D., and Terzi, C. (eds.) (2012). *L'expérience des problèmes publics*. Aubervilliers: Éditions de l'École des hautes études en sciences sociales.

Central Statistics Office. (2017). *Regional SDGs Ireland 2017*. https://www.cso.ie/en/release sandpublications/ep/prsdgi/regionalsdgsireland2017/env/.

Corr, S. (2020, November 28). Amazon staying tight lipped on landslide at wind farm construction site. *Belfast Live*. https://www.belfastlive.co.uk/news/belfast-news/amazon-staying-tight-lipped-landslide-19312168.

Couldry, N., and Mejias, U. A. (2019). Data colonialism: rethinking Big Data's relation to the contemporary subject. *Television and new media*, 20(4), 336–49. https://doi.org/10. 1177/1527476418796632.

Crawford, K. (2021). *Atlas of AI: power, politics, and the planetary costs of artificial intelligence*. New Haven, CT: Yale University Press.

Daggett, C. N. (2019). *The birth of energy: fossil fuels, thermodynamics and the politics of work*. Durham, NC: Duke University Press.

Danish Energy Agency. (2018). *Analysis of hyperscale data centres in Denmark*. https://ens. dk/sites/ens.dk/files/Analyser/analysis_of_hyperscale_datacentres_in_denmark_-_eng lish_summary_report.pdf

Debruyne, O. (2020, January 23). Saturé par les Data centers, le réseau électrique d'Île-de-France va-t-il craquer? *Le Parisien*. https://www.leparisien.fr/seine-saint-denis-93/sature-par-les-data-centers-le-reseau-electrique-d-ile-de-france-va-t-il-craquer-23-01-2020-8242308.php.

Dewey, J. (1927). *The public and its problems*. New York: Henry Holt and Company.

Dourish, P. (2017). *The stuff of bits: an essay on the materialities of information*. Cambridge, MA: MIT Press.

Dupont-Calbot, J. (2013, May 27). La Seine–Saint-Denis, "Data Valley" du *cloud* français. *Le Monde*. https://www.lemonde.fr/a-la-une/article/2013/05/27/la-seine-saint-denis-data-valley-du-cloud-francais_3418227_3208.html.

Emerson, R. M., and Messinger, S. L. (1977). The micro-politics of trouble. *Social problems*, 25(2), 121–134.

Eurostat (2022, January 26). *Renewable energy on the rise: 37% of EU's electricity*. Retrieved 24 February 2022, from https://ec.europa.eu/eurostat/fr/web/products-eurostat-news/-/ ddn-20220126-1

Gabrys, J. (2013). *Digital rubbish: a natural history of electronics*. Ann Arbor: University of Michigan Press.

Gillis, J. (2004). *Islands of the mind: how the human imagination created the Atlantic world*. New York: Palgrave Macmillan.

Gilmore, J. N., and Troutman, B. (2020). Articulating infrastructure to water: agri-culture and Google's South Carolina data center. *International journal of cultural studies*, 23(6), 1–16.

Government of Iceland (n.d.). *Ministry of industries and innovation, energy.* Government of Iceland. Retrieved January 19, 2022, from https://www.government.is/topics/business-and-industry/energy/

Graham, S. (2016). *Vertical: the city from satellites to bunkers.* London: Verso Books.

Graham, S., and McFarlane, C. (eds.) (2015). *Infrastructural lives: urban infrastructure in context.* Routledge, Taylor, and Francis Group.

Greenhough, B. (2006). Tales of an island-laboratory: defining the field in geography and science studies. *Transactions of the Institute of British Geographers*, 31(2), 224–237.

Guðmundsson, S. (2019). Hvalá fyrir bitcoin? [Hvalá river for Bitcoin?]. *Kjarninn*, 14 May. https://kjarninn.is/skodun/2019-05-14-hvala-fyrir-bitcoin/

Guma, P. K. (2019). Smart urbanism? ICTs for water and electricity supply in Nairobi. *Urban studies*, 56(11), 2333–2352.

Hogan, M. (2015). Data flows and water woes: the Utah data centre. *Big Data and society*, 2(2), July–December, 1–12.

Hogan, M. (2018). Big Data ecologies. *Ephemera: theory and politics in organisation*, 18(3), 631–657.

Holt, J., and Vonderau, P. (2015). 'Where the Internet lives': data centres as cloud infrastructure. In L. Parks and N. Starosielski (eds.), *Signal traffic: critical studies of media infrastructures.* Urbana: University of Illinois Press. 71–93.

Høvsgaard Nielsen, L. (2014). *(Re)configuring green data storage: an ethnographic exploration of an industry in the making.* Masters thesis, IT University of Copenhagen.

Howe, C., et al. (2015). Paradoxical infrastructures. Ruins, retrofit, and risk. *Science, technology and human values.* http://doi.org/10.1177/0162243915620017

Hu, T.-H. (2015). *A prehistory of the cloud.* Cambridge, MA: MIT Press.

Iceland Chamber of Commerce. (2019). *The Icelandic economy: current state, recent developments and future outlook.* https://www.vi.is/files/%C3%BAtg%C3%A1fa/sk%C3%BDrslur/the_icelandic_economy_2019_report.pdf.

Ingólfsson, B.B. (2019, July 16). Gagnaver nota jafnmikla orku og heimilin [Data Centers Use as Much Energy as Homes]. *RÚV.* https://www.ruv.is/frett/gagnaver-nota-jafnmikla-orku-og-heimilin

Irish Peatland Conservation Council (2020). *Climate Change and Irish Peatlands.* Retrieved 19 January 2022, from: https://www.ipcc.ie/a-to-z-peatlands/irelands-peatland-con servation-action-plan/peatland-action-plan/climate-change-and-irish-peatlands/.

Ísleifsson, S. (2011). Islands on the edge: medieval and early modern national images of Iceland and Greenland. In S. Ísleifsson (ed.), *Iceland and images of the North.* Sainte-Foy: Presses de l'Université de Québec. 41–66.

Jevons, W. S. (1865). *The coal question: an inquiry concerning the progress of the nation, and the probable exhaustion of our coal-mines.* London: Macmillan.

Johnson, A. (2019). Data centers as infrastructural inbetweens. *American ethnologist*, 46, 75–88.

Jones, Nicola. 2018. How to stop data centres from gobbling up the world's electricity. *Nature*, 561, 163–167.

Judge, P. (2020, July 1). Amsterdam resumes data centre building, after a year's moratorium. *Data center dynamics.* https://www.datacenterdynamics.com/en/news/amsterdam-re sumes-data-center-building-after-years-moratorium/

Judge, P. (2021, March 13). Frankfurt to regulate data centres. *Data Center Dynamics.* https:// www.datacenterdynamics.com/en/news/frankfurt-to-regulate-data-centers/

Kiernan, A. (2019, April 11). Donegal group 'baffled' by Amazon deal to wind farm awaiting planning permission. *National wind watch.* https://www.wind-watch.org/news/2019/04/ 11/donegal-group-baffled-by-amazon-deal-to-wind-farm-awaiting-planning-permission/

Lally, N., Kay, K., and Thatcher, J. (2019). Computational parasites and hydropower: a political ecology of bitcoin mining on the Columbia River. *Environment and planning E: nature and space,* 1–21.

Larkin, B. (2013). The politics and poetics of infrastructure. *Annual review of anthropology,* 42(1), 327–343. https://doi.org/10.1146/annurev-anthro-092412-155522.

Leicher, D. (2013). *Les data centres sur Plaine Commune.* Agence Locale de l'Énergie et du Climat de Plaine Commune. https://www.alec-plaineco.org/IMG/pdf/alec-plaine-com mune-2013-les-data-centers-sur-plaine-commune.pdf.

Le Parisien. (2011, April 14). La pénurie d'énergie menace les data centres [Press release]. https://www.leparisien.fr/seine-saint-denis-93/la-penurie-d-energie-menace-les-data-cen ters-14-04-2011-1406697.php.

Libertson, F., Velkova, J., and Palm, J. (2021). Data-centre infrastructure and energy gentrification: perspectives from Sweden. *Sustainability: science, practice and policy,* 17(1), 153–162.

Liboiron, M., and Lepawsky, J. (2022). *Discard studies: wasting, systems, and power.* Cambridge, MA: MIT Press.

Light, J. S. (1999). When computers were women. *Technology and culture,* 40(3), 455–483.

Lohmann, L. (2016). What is the 'green' in 'green growth'? In G. Dale, M. Mathai, and J. Puppim de Olivera (eds.), *Green growth: ideology, political economy and the alternatives.* London: Zed Books. 42–71.

Magnason, A. S. (2006). *Dreamland: a self-help manual for a frightened nation.* Rochester, NY: Open Letter Books.

Magnússon, J. B. (2015, September 6). Vafasamar tengingar stærsta gagnavers landsins [Suspicious connections at the nation's largest data center]. *Stundin.* https://stundin.is/ frett/vafasamar-tengingar-staersta-gagnavers-landsins/

Maguire, J. (2020). Icelandic resource landscapes and the state: experiments in energy, capital, and aluminum. *Anthropological journal of European cultures,* 29(1), 20–41.

Maguire, J., and Winthereik, B. R. (2021). Digitalizing the state: data centres and the power of exchange. *Ethnos,* 86(3), 530–551. https://doi.org/10.1080/00141844.2019.1660391.

Maguire, J., Watts L., and Winthereik, B. R. (eds.) (2021). *Energy worlds in experiment.* Manchester: Mattering Press.

Masanet, E., Shehabi, A., and Lei, N. (2020). Recalibrating global data center energy-use estimates. *Science,* 367(6481), 984–86. https://doi.org/10.1126/science.aba3758.

Mattern, S. (2017). *Code and clay, data and dirt: five thousand years of urban media.* Minneapolis: University of Minnesota Press.

Mayer, V. (2021). From peat to google power: communications infrastructures and structures of feeling in Groningen. *European journal of cultural studies,* 24(4), 901–915.

McSweeney, E. (2021, February 11). The science of bogslides: we must learn how to judge the risks. *The Irish Times*. https://www.irishtimes.com/news/science/the-science-of-bog slides-we-must-learn-how-to-judge-the-risks-1.4468422.

Oslund, K. (2011). *Iceland imagined: nature, culture, and storytelling in the North Atlantic*. Seattle: University of Washington Press.

Pasek, A. (2019). Managing carbon and data flows: fungible forms of mediation in the cloud. *Culture machine*, 18. http://culturemachine.net/vol-18-the-nature-of-data-centers/manag ing-carbon/.

Perón, F. (2004). The contemporary lure of the island. *Tijdschrift voor economische en sociale geografie*, 95(3), 326–39.

Pickren, G. (2017). The factories of the past are turning into the data centers of the future. *Imaginations: journal of cross-cultural image studies*, 8(2), 22–29.

Pollio, A., and Cirolia, L. R. (2022a). *Financing ICT and digitalisation in Africa: Current trends and key sustainability issues*. Cape Town: African Centre for Cities and Alfred Herrhausen Gesellschaft.

Pollio, A., and Cirolia, L. R. (2022b). Fintech urbanism in the startup capital of Africa. *Journal of cultural economy*. https://doi.org/10.1080/17530350.2022.2058058

Quann, J. (2019, April 8). Donegal wind farm to power Amazon infrastructure. *Newstalk*. https://www.newstalk.com/news/donegal-wind-farm-power-amazon-infrastructure-845990.

Service des études financières (2012, November 15). *Note sur la fiscalité (CET) des datacenters sur le territoire de Plaine Commune*. Département administration générale, finances, évaluations et outils de pilotage de Plaine Commune.

Skúlason, J., and Hayter, R. (1998). Industrial location as a bargain: Iceland and the aluminum multinationals 1962–1994. *Geografiska annaler, Series B, Human geography*, 80(1), 29–48.

Star, S. L., and Ruhleder, K. (1996). Steps towards an ecology of infrastructure: design and access for large information spaces. *Information systems research*, 7(1), 111–134.

Starosielski, N. (2015a). *The undersea network*. Durham, NC: Duke University Press.

Starosielski, N. (2015b). Fixed flow: undersea cables as media infrastructure. In L. Parks and N. Starosielski (eds.), *Signal traffic: critical studies of media infrastructures*. Urbana: University of Illinois Press: 53–70.

Stefánsdóttir, B. (2019, September 19). Fjögur gagnaver orðin stórnotendur raforku [Four Data Centers Become Heavy Users of Electricity]. *Kjarninn*. https://kjarninn.is/frettir/2019-09-06-fjogur-gagnaver-ordin-stornotendur-raforku/.

Strauss, S., Rupp, S., and Love, T. (eds.). (2013). *Cultures of energy: power, practices, technologies*. Walnut Creek, CA: Left Coast Press.

Swinhoe, D. (2021, January 17). Bill banning new data centre developments introduced in Ireland. *Data center dynamics*. https://www.datacenterdynamics.com/en/news/bill-ban ning-new-data-center-developments-introduced-in-ireland/.

Taffel, S. (2021). Data and oil: metaphor, materiality and metabolic rifts. *New media and society*, 1–19. https://doi.org/10.1177/14614448211017887.

Taylor, A. R. E. (2018). Failover architectures: the infrastructural excess of the data centre industry. *Failed architecture*. https://failedarchitecture.com/failover-architectures-the-in frastructural-excess-of-the-data-centre-industry/.

Taylor, A. R. E. (2019). The data center as technological wilderness. *Culture machine*, 18, 1–30.

Taylor, A. R. E. (2021a). Future-proof: bunkered data centres and the selling of ultra-secure cloud storage. *Journal of the Royal Anthropological Institute*, 26(S1), 76–94.

Taylor, A. R. E. (2021b). Standing by for data loss: failure, preparedness and the cloud. *Ephemera: theory and politics in organisation*, 21(1).

Taylor, A. R. E. (2022a). Bunkers, data, preparedness: from the mushroom cloud to the computing cloud. *New media and society*, online first, 1–25.

Taylor, A. R. E. (2022b) Cloud labour: data centre work and the maintenance of media infrastructure. *The Routledge companion to media anthropology.* London and New York: Routledge.

Velkova, J. (2016). Data that warms: waste heat, infrastructural convergence and the computation traffic commodity. *Big Data and society*, 1–16. doi: 10.1177/2053951716684144.

Velkova, J. (2019). Data centres as impermanent infrastructures. *Culture machine*, 18.

Velkova, J. (2021). Thermopolitics of data: cloud infrastructures and energy futures. *Cultural studies*, 35(4–5), 663–683. https://doi.org/10.1080/09502386.2021.1895243.

Vemuri, A. (2022). Fieldwork in Covid times. *Grierson Research Group.* Retrieved 22 January 2022, from https://www.griersonresearchgroup.ca/field/covid-times.

Vonderau, A. (2018). Technologies of imagination: locating the cloud in Sweden's global north. *Imaginations: journal of cross cultural image studies*, 8(2), 8–21.

Vonderau, A. (2019a). Storing data, infrastructuring the air: thermocultures of the cloud. *Culture machine*, 18.

Vonderau, A. (2019b). Scaling the cloud: making state and infrastructure in Sweden. *Ethnos*, 84(4), 698–718. https://doi.org/10.1080/00141844.2018.1471513.

von Schnitzler, A. (2016). *Democracy's infrastructure: techno-politics and protest after apartheid.* Princeton: University Press.

Watts, L. (2018). *Energy at the end of the world: An Orkney Islands saga.* Cambridge, MA: MIT Press.

Karen Waltorp, Ragnhild Freng Dale, Martín Fonck, and Pierre du Plessis

5 Imagining energy futures beyond colonial continuation

Introduction

The enduring material effects that colonial extractivism has had on human and more-than-human life, and on geopolitical formations, continue to shape imaginaries of future energy security in new ways. This chapter interrogates this context through an examination of (de)colonial energy futures through empirically grounded investigation. By bracketing '(de-)' in 'decolonial', we wish to stress both how continuities of colonial imaginings of energies and landscapes unfold and how they could be imagined otherwise in paying attention to local struggles and alternatives. The colonial legacies in perceiving the environments we inhabit as spaces to be conquered or as 'standing reserves' (Heidegger 1977, see Waltorp et al. forthcoming) resonate into the present, particularly in the way such treatments have become central to capitalist modes of extraction and expansion. This, however, is only one of many possible framings, and it interacts in unforeseen ways with alternative ecological imaginings of landscapes as relational forces and energies – from below the earth's surface, to water and ice and into the air.

Colonial continuation

Resource prospecting and extraction of energy were driving forces of colonial expansions. The dominant period of European colonialism and colonial occupation officially came to an end in the twentieth century as many (though not all) colonies and protectorates gained independence. As current, ongoing processes and calls for decolonization draw attention to, it is clear that colonial projects have been rearticulated in economic terms, emerging as new scrambles for *power* – material and political over the 'standing reserves' across the globe. The growing calls for decolonising both within and outside the academy, indicated by the great number of post- affixes in the academic literature, reflects this state and co-existing processes: There cannot be a post-colonial, when the colonial project never ended (Kelley 2000: 27). Decolonial scholarship has focused on epistemological and ontological debates, as well as on getting practical.

https://doi.org/10.1515/9783110745641-006

Imagination and the capacity to imagine otherwise (Escobar 2007, McTighe and Raschig 2019) is part of this; a fruitful field of analysis, contestation, and possibilities.

Scholars whose thinking and writing inspire our understanding of decoloniality and a decolonial practice point to the importance of the simultaneous decolonization of countries – of land *and* of minds. We draw from various traditions. Martinique-born Franz Fanon, writing from French-occupied Algeria, is often counted as the main reference with his now classic works Black skin, white masks (2008), The wretched of the earth (1963), and A dying colonialism (1967). With other Caribbean scholars, Aimé Césaire (2000), Stuart Hall and colleagues (1996), and Èdouard Glissant (1997), this is a starting point: Césaire writes in *Discourse on Colonialism* (in French, 1950) that a way forward is neither to follow Europe's footsteps and a 'master classes' ideology of progress, nor to go back to ancient ways. It is carving out a decidedly new way, imagining futures with hybrid pasts and beyond colonial (double-)alienation (Césaire 2000). Hall points to the need for dynamic *articulations* (always full of friction) that allow us to enter into temporary constellations with *others* and forging (co-)articulations with overlapping stakes (Hall, Morley, and Chen 1996, see also Waltorp and ARTlife Film Collective 2021); Kim TallBear (2013) looks at the entanglements of land, minds, and genes in articulations of identity and belonging and rights within dominant US discourse. Other contemporary voices ask what the case is for simply letting anthropology burn? To start over elsewhere (Jobson 2020, see also Allen and Jobson 2016 for an overview). Another moment of inspiration in the 1980s counted Chandra Mohanty (1988), Gayatri Spivak (1988), Trinh T. Minh-ha (1990) and other feminist and subaltern studies scholars, who have inspired the discipline of anthropology.[1]

In South Africa a number of scholars contribute to the development of thinking and practising a decolonial scholarship/anthropology that is underlying the approach we forward here; from Steve Biko (2002/1978), to Bernard Magubane (1979), and up to Zimitri Erasmus (2017) and colleagues. From the perspective of a Sámi indigenous scholar, Rauna Kuokkanen, echoing Césaire, we should be wary of the tendency to represent Indigenous culture as static and made 'impure' by colonisation (Kuokkanen 2000). In Sápmi, it is often artists who play leading parts in political mobilisation, using artworks as decolonial practice and counternarratives (Stephansen 2017, Sandström 2020). Decolonisation is

---

1 In their foreword to the second edition of *Anthropology as cultural critique* (1991), George Marcus and Michael Fisher point to their omission in the first edition in 1986 of the fundamental inspiration growing out of feminist anthropology and theory broadly speaking for the discipline at that moment.

also an everyday practice of recovering from fragmentation and language loss (Dankertsen 2016). In this vein, with the above tradition(s) as inspirations, what is the relevance of analysing imagination and energy narratives to produce futures otherwise? As mentioned in the Introduction to this book, existing scenario planning and forecasting used in the energy sector often tend to prioritise and limit possible futures so that 'the overall framework of energy transitions has narrowed the scope of how anthropologists understand and engage in ethical dilemmas posed by energy' (High and Smith 2019: 11). Understanding these dilemmas therefore presents new challenges for anthropology to include the questions, desires, and concerns of groups of people living in separate but interdependent worlds. We go beyond 'just' a critique of existing narratives, visions, and research to articulate this as an interventional agenda towards equitable futures that, as Donna Haraway points out, must be 'friendly to earthwide projects of finite freedom, adequate material abundance, modest meaning in suffering, and limited happiness [...] insistence on irreducible difference and radical multiplicity of local knowledges [...] an earthwide network of connections, including the ability partially to translate knowledges among different – and power-differentiated – communities' (Haraway 1988: 579–580).

The ethnographic examples in this chapter explore the simultaneous confrontational strategies, the battles, and the ongoing everyday implications of differently situated actors. These frictions must be understood in relation to the backdrop of energy paradigms located in specific and differentiated colonial pasts, and their relational configurations of energy infrastructures still in the present. In developing this discussion the chapter traces how energy futures are contested today, often 'out of sight' but continually glimpsed by the public through research, art, and activism in Sápmi and the Norwegian North (Dale); over the possible futures imagined by engineers and geologists interacting with geothermal energy in the the Chilean Andes (Fonck); in tracking natural gas prospectors in Botswana's Kalahari Desert, and a different paradigm of environmental attunement for understanding the consequences of natural resource 'speculation' (du Plessis); and finally the fraught electricity situation, use of mobile phone chargers, and expensive data in the impoverished Cape Flats township area in South Africa, where coal cannot any longer sustain the energy demand (Waltorp). Politicians place hopes in 'green' transitions(s) and there is significant movement towards cleaner, more renewable energy. In Northern Norway, however, 'green' energy alternatives promoted threaten traditional Sámi livelihoods; and in Botswana crucially important wildlife areas in remote areas of the Kalahari, home to indigenous San and Bakgalagadi communities, become sites for speculative energy futures as the state seeks to shift away from energy dependence on neighbouring South Africa. South Africa pins its hopes on wind energy, while in Chile 'green' transitions are reproducing the environmental impacts of a broader logic of extraction, opening new frontiers for speculation.

Our starting point for this chapter is that any discussion about future energies is necessarily entangled with, and should be attentive to and wary of, the violent histories and colonial legacies of human exploitation and environmental devastation associated with the pursuit of material and political power that resonate into the present. As our work engages with minority and/or Indigenous populations in different ways, our positions as anthropologists at European institutions situated in the Global North must also be acknowledged – and is grappled with in different ways through our empirical examples below. We do not claim to speak on behalf of our interlocutors. Rather we situate our anthropological writing and other modes of action in dialogue with the pasts, presents, and futures we research. We also take inspiration from the strands of decolonial thinking and practice referred to above. This leads us to 'get practical and imaginative', to ask: ultimately, how can anthropology contribute to imagining and practising futures otherwise? The 'otherwise' has been understood and felt to enjoin scholars to an enduring struggle for liberation within fields such as Black, Indigenous, Latinx, Asian American, postcolonial, queer, and gender studies (McTighe and Raschig 2019). With firm foundations in social movements, the otherwise 'summons simultaneously the forms of life that have been able to persist despite constant and lethal forms of surveillance, as well as the possibility for, even the necessity of, abolishing the current order and living into radical transformations of worlds' (McTighe and Raschig 2019). Arturo Escobar's (2007) position on anthropologists' position within 'development' as discourse, idea, and billion-dollar industry in a neoliberal and developmentalist capitalist modernity saw him propose radical, experimental, and exploratory alternatives to development located in community and indigenous knowledge systems, in intentional alternative lifeways, as well as the civil society activities of organised social movements:

> ecology and environmentalism imply different ways of thinking (necessarily relational, situated and historical); ways of reading modernity; an acute concern with epistemology (particularly a critique of reductionist science and logocentric discourse); and an articulation of the question of difference (ecological and cultural difference) that can easily be linked to coloniality, and vice versa (Escobar 2007: 196–197).

Which different ways of imagining futures emerge from paying attention to the situated and material aspect of energy? What are the spaces of (im)possibility of imagination in these contexts? Following where the energy comes from and the (in)visibility of energy infrastructure, is also a way of studying the complexity of emerging narratives from a situated perspective (Abram, Winthereik, and Yarrow 2019, Anand, Gupta, and Appel 2018, Larkin 2013, Bowker and Star 1999) opening up to imagining energy futures otherwise.

In this regard, what is the relevance of analysing imagination and energy narratives to produce futures otherwise? In imagining – or more tenuously, pursuing – energy futures, it might serve us well to be reminded that 'Extractive injustice and mass extinctions are exacerbated by unfettered expansion' (Strathern et al. 2019). That is, as energy sources, resources, and extractive practices are reconfigured and reconceptualized to better respond to the challenges of the Anthropocene and working towards more livable futures, these ambitions should be careful not to reproduce those practices and logics of 'unfettered expansion', so foundational to colonialism and colonial exploitation. Such expansion is a material enactment of market logics of capitalist growth-without-limits, which, as Julie Livingston (2019) puts it, has led us into the predicament of 'self-devouring growth', where endless growth is literally destroying the material conditions on which growth itself depends.

Imagining energy futures

Thinking that takes futures into account is now present in the majority of anthropological energy research. Studying energy futures is an emerging field that opens analytical and practical possibilities for imagining and practising futures otherwise, forging new forms of collaboration and producing 'spaces of possibilities' through which descriptions of partial (im)possibilities can be generated and embraced, while escaping the temptation of totalizing diagnostic discourses and reductionist solutionism. These spaces of possibility highlight the diverse dimensions of the challenges we face. Specific social, cultural, and material conditions allow groups to imagine futures otherwise, and it also 'matters what senses of futurity we bring into play' (Wilkie, Savransky, and Rosengarten 2017). In this regard, the ethnographic register can make a contribution in analysing how energy futures are performed in specific contexts (Watts 2018), which logics are reproduced, and which ones can be challenged.

In this chapter we present four ethnographic cases. The first interrogates how an arts movement and artwork illuminates the way futures are shaped by colonial legacies and contemporary imaginaries about energy infrastructure and resource extraction, and also intervenes to change this course of events. The Pile o'Sápmi project and collaborative advocacy described below demonstrates very clearly how an artwork makes possible conversations that normally are silenced and makes visible colonial practices that continue to manifest today.

This is important to the anthropological community because it is about our rela-tionships in and outside the field, and it is important to Ragnhild Freng Dale's interlocutors because it is a struggle for the right to exist, to a wider international context of indigeneity and coloniality. And not least, it is important to Norwe-gian majority society as there is too little knowledge about the country's colonial legacy.

The second ethnographic case highlights the role of non-human elements, such as underground water and geysers in energy futures imagination. Specifi-cally, Martín Fonck focuses on geothermal energy exploration in the geothermal field El Tatio in the Chilean Andes from a historical perspective. Following how these futures are made through exploratory drilling wells and underground water speculation, this chapter explores the challenges of making visible non-human assemblages as a potential source of electric energy. Analysing drilling practices as continuation of colonial logics, this chapter also discusses the un-finished nature of energy futures.

What follows in the third ethnographic case also relates to water. Pierre du Plessis elaborates on how resource prospecting that occurs in remote regions builds its momentum slowly and out of sight, in a bumbling manner. The as-sumption seems to be that such areas are simultaneously underutilised and vul-nerable, and therefore full of profitable extractivist potential. du Plessis explores a specific case in which those assumptions are enacted through practices of speculation and prospecting for Coal Bed Methane (CBM) 'reserves' in a remote region of the Kalahari, Botswana in ways that significantly affect water resources for residents in the region. The bumbling involved in prospecting secures finan-cial investment and futures through the prospect of striking it rich and potential-ly securing energy futures. But it is a messy process, occurring at a distance with relatively little attention, and often requires sidestepping competing ontologies of a given landscape, such as those that enact environmental conservation, wild-life reserves, and even the commons as they relate to remote area dwellers, as Kalahari communities are often referred. In doing so, resource frontiers are enacted instead as 'standing reserves', long before the first drills even make con-tact with the strata and the liveness of landscapes begin to push back.

The fourth ethnographic case is geographically close to Pierre du Plessis's field site. The empirical case which Karen Waltorp introduces sheds light on how intimate relations, energy, cables, data, and money become entangled in Manenberg – a predominantly 'Coloured' township on the outskirts of Cape Town. The so-called 'Coloured' population was categorised as such preceding apartheid rule (1948–1994), but during this era's obsession with perceived 'race' purity, all people who in accordance with this logic were considered 'mixed-race' were classified thus. A category in-between that of the privileged

'White' rulers in power who benefited economically from this system, and the 'Black African' majority with less opportunity and fewer rights afforded them. Your assigned 'race category' determined where you could live, whom you could marry, what education was available to you, and what job you could take up – even which beach you could go to and which bench you could sit on in public space (Adhikari 2006, Erasmus 2017). This diverse group is a minority in South Africa, where 8.9 percent of the population identify as 'Coloured' according to the National Census 2011 (Stats SA 2012). Many today identify as 'Coloured' and find pride in all that was and continues to be a part of that diverse experience and identity beyond the apartheid regime, including Indigenous ancestry. Others have adopted the term 'so-called Coloured' and/or identify as black – rather as political identity and not the apartheid classification 'Black' with a capital B, as they also reject the term 'Coloured' and with it the supremacist ideology underlying its use.

Mobile devices are critical as relational devices in connecting people to each other and to various services in Manenberg – far removed from the centre and commerce due to apartheid's geography – as elsewhere in today's world. Prepaid mobile phones *need* data (also known as airtime, credit, or talktime) so that users can make calls, send texts, and access online services. The energy infrastructure as well as the telecommunications infrastructure of fibre optic cables makes this quite complicated, not least so for inhabitants in socio-economically, historically deprived areas such as Manenberg. In all the places which the four case studies focus on, the past is contested and friction emerges around futures possible depending on how the past is (re)-claimed after decades of dispossession. Energy futures are at the heart of this.

We begin in the North.

Art interventions and coloniality in the Norwegian north

Ragnhild Freng Dale

Oslo, Norway, December 2017

Somewhere between the Google documents we are working on and the fourth or fifth batch of coffee from a cafetière, a programme for an artwork and manifestation beginning in a few days is starting to come together. The front page is a shade of eerie green, a colour which spills onto the reindeer skulls with bullet holes through their forehead. The words on the page read 'Pile o'Sápmi Supreme' in white and 'There is no postcolonial' in red letters, contrasting to the cold, photographed skulls. I am sitting with the artist Máret Ánne Sara, the lead

organiser and artist behind the event. The programme we are proofreading explains the con-
text of the Pile o'Sápmi-project – an arts movement as well as a title work, aiming to highlight
and create debate about the current issues and colonial practices that affect the lives of the
Sámi; not in a distant past, but today, here and now, through the legal structures of the Nor-
wegian nation-state.

On a December morning of 2017, the young reindeer herder Jovsset Ánte Sara entered the Supreme Court in Norway, with the Norwegian government as his op-
ponent. He had sued the state to demand his right to continue herding reindeer,
a future threatened by their demand he reduce his herd to such a small number
of animals he would be forced to quit herding altogether. Sara won in both the
District Court and the Court of Appeal, but the state had appealed yet again, and
his fate was now in the Supreme Court's hands. To support him and draw attention to the case, his sister and artist Máret Ánne Sara created the work *Pile o'*
Sápmi before the first court hearing in 2016, and invited other artists to contribute. The initiative grew into a wider movement that made visible the injustices
Sámi reindeer herders faced, and when I joined before the Court of Appeal hear-
ings in Tromsø, there were already a wide pool of contributors; Sámi, non-Sámi,
artists, activists, academics, and others. In Tromsø, Máret Ánne had invited artists from across Sápmi to participate in an ambitious programme of art events,
talks and exhibitions hosted by the art gallery Small Projects and other venues.
The streets were filled with the work of the anonymous art collective Suophan-
terror, as well as street theatre and a graffiti work by Anders Sunna and Linda
Zina Aslaksen, depicting the lion in the Norwegian coat-of-arms grabbed by its
tail by a reindeer skeleton, whose arms are about to cut it off.

I had come to know the Sara siblings during my fieldwork in Northern Norway and Sápmi, as I was researching how the local community in the town of
Hammerfest lived with the experience of becoming a petroleum town (Dale
2018). I didn't then know how governmental regulations and new legislation
on reindeer herding were impacting herders whose summer pastures are in
the same area, but Máret Ánne Sara's visual art spoke in a language that
made their impacts very clear, with graphic representations that tore off the
clean violence of the bureaucracy through which the Norwegian nation state gov-
erns.

The following ethnographic case interrogates how art practices can carve
open a space to re-envisage the present and what is deemed possible and fair
futures. It grows out of my ongoing reflection on the relationship between my
work as an ally with Sámi artists and activists, and my practice as an anthropologist trained in a conservative UK institution. Conventional academic work is,
I have found, too slow in the face of injustices I encountered in the field, its de-

Fig. 5.1: Graffiti by Anders Sunna and Linda Zina Aslaksen. Photo by Máret Ánne Sara.

liberations post-event useless to my interlocutors: if the reindeer are slaughtered and there is no more future to be held open, no amount of analysis or critique can reverse it. A research project which doesn't grapple with the ways in which the researcher either supports or works to disentangle the violence of such structures fails to see the liberatory potential that *is* embedded in anthropology, and which has come out through a number of engaged anthropologies since the beginning of the discipline. Starting from the artwork and its many iterations, I seek to discuss how this kind of arts practice manifests alternative futures, and how anthropological interventions, sometimes under the radar, might be constructive (or less constructive) contributions to keep those possible futures open.

Making reindeer skulls speak

Pile o' *Sápmi*, both an artwork by Máret Ánne Sara and an artist-led movement with the same name, has by now travelled the world, from outside the Inner Finnmark District Court in Deatnu, the Arctic Arts Festival in Harstad, to the Court of Appeal in Tromsø, the arts festival Documenta 14 in Athens, Greece and Kassel, Germany, and now to Oslo, the capital of Norway. Her use of the reindeer skulls as artwork(s) were a direct reaction to the violence performed by bureaucratic structures. During the December night which introduced this chapter, the skulls were about to be installed outside the Norwegian Parliament, in order to greet parliamentary politicians and the public in Oslo the morning of the Supreme Court hearing a few blocks away. Present were also activists and artists, politicians, and other supporters. They sang, *joiked*, and gave short speeches for Jovsset Ánte Sara and the right of his reindeer herd to survive.

The skulls were the latest version of a visceral artistic statement. Their first incarnation was a bloody pile of reindeer heads outside the Inner Finnmark District Court in Deautnu in 2016, with a Norwegian flag planted on top. This picture circulated widely in the media, and became a chance for Máret Ánne to speak about the reason for the artwork: the plight of her little brother, Jovsset Ánte Sara, who had sued the Norwegian government to protect his cultural rights and prevent the implementation of the forced culling. The culling was based on the notion from the government that there are too many reindeer in the districts whose summer pastures lie in West Finnmark. A remaking of the law was implemented in 2007, and enabled forced culling if the herders failed to comply with the maximum amount of reindeer the government decided for each district. The new law continues the rationalisation of Sámi reindeer herding by the government since the 1970s – and the methods on which it is based has been thoroughly critiqued by scholars from political ecology and ethnology (Bjørklund 2016, Johnsen, Benjaminsen, and Eira 2015). Bureaucratically, the measures were called 'forced reduction' policies (*tvangsreduksjon*), but it has more widely become known as forced slaughter (*tvangsslakt*), as the effect of reduction is the involuntary slaughter of animals that would otherwise have lived on as members of the reindeer herd.

In Sara's district, where he is one of six rights holders, the government had set a limit of 2,000 reindeer, but only four out of six herders had come to a mutual agreement on who should reduce by how much. Though these four had allowed the last third to retain a third of the reindeer numbers, the government demanded it was a unanimous decision or none, and ordered all of the herders should reduce by a certain percentage. This 'proportional' reduction was made without any special arrangements to protect those with the fewest animals, even though such shielding was demanded by both the Sámi Parliament and the Norwegian Reindeer Herding Association (NRL) when the law was passed in 2007. As a consequence, Sara was ordered to reduce his already small herd to 75 animals – a number that would effectively put him out of his livelihood and force him out of herding, as such a herd would not be financially viable or resilient enough in size in case of a bad year or attacks by predators. Sara took the case to court, claiming the Norwegian government's decision breached his human rights and his rights as an Indigenous person to practise his culture, rooted in the European Charter of Human Rights and the UN Convention on Indigenous Peoples. Sara won in the District Court and the Court of Appeal, but the state appealed to the Supreme Court in Oslo. Unsurprisingly, the Sara siblings were worried that the stakes for the state were too high for him to win. Should the judges rule that the government's order denies Sara his human rights to prac-tise his culture, then the entire law might need to be rewritten. A few weeks later the outcome of the verdict confirmed their worries: though the court thinks there *is* a right to practise his culture, the protection of the grazing land from overgraz-ing took priority.

Our work in Oslo was to make sure the court case was visible to a wider pub-lic outside the courtroom, that the story was told, and the perspectives of Sara and his allies were heard across different media channels. It also carried signifi-cant symbolic value to do this in front of Parliament, following in the footsteps of Sámi activists who had occupied the same space 40 years earlier, when they fought for recognition and to stop the damming of the Alta river in Finnmark. Though much was won at that time, including the official recognition of the Sámi as indigenous people and the establishment of the Sámi Parliament in 1989, Sámi issues have fallen off the agenda in Norwegian majority society, where there is little knowledge about the situation today – issues mostly met with ignorance or prejudice. To bring them back into the consciousness of people in Oslo, one of the contributing artists, A. K. Dolven, invited volunteers to carry a skull around the city for a day during the court proceedings, to talk to strangers about what was going on. Anthropologist Hugo Reinert describes them in the fol-lowing manner:

> Flowing out from their point of release they enact an uncanny transformation, reshaping the city by revealing it as it already was: a landscape of colonial violence made stone, haunted by the exclusions it simultaneously effects and depends on. (Reinert 2019).

A. K. Dolven herself, who as a child had experienced the struggles for the Alta river, further asked why Norway has scraped away the true story of its In-digenous people, just as Máret Ánne's skulls are without a trace of flesh. The skulls, with their eerie yet beautiful presence, invited conversations and spaces of learning, acts of solidarity and action. A few of us brought a skull with us into the Supreme Court, passed through the security check, and held them as we sat on the benches open to the public: reindeer skulls witnessing the highest in-stance of the Norwegian law, as they deliberated whether or not the living flesh of Sara's herd would come to the same fate. Sara himself was present, dressed in the traditional *gákti* from his home region. Other Sámi allies had also put on their gáktis, making it visible that many Sámi were there to witness the proceedings.

Fig. 5.2: Detail of *Pile o'Sápmi Supreme* hanging in front of the Norwegian Parliament. Photo by Ragnhild Freng Dale.

The flow of the skulls even extended into international headlines: 'In Norway, Fighting the Culling of Reindeer With a Macabre Display', read the heading in

Fig. 5.3: A skull held outside the entrance to the Supreme Court. Photo by Inga Marie Nymo Riseth.

the *New York Times* on December 6, the artwork the clear entry point into an extensive reportage of the legal case. Speaking to the international press, Máret

Ánne Sara stated her frustration with not being heard in the majority society. The picture was reprinted in several newspapers across the world, as a rather unflattering expression of how Norway treats its Indigenous population. Following Reinert's observation (2019), the spectres of the once-living reindeer are now at work, acting like a weapon in the fight to keep Jovsset Ánte Sara's herd alive and to dismantle the structures that stand in the way of this future.

Enunciatory collaboration

The frustration of the Sara siblings might, at first sight, seem disconnected from matters concerning energy, and better understood as a matter of pasture lands and sustainability. But the government's concern with a maximum number of reindeer for sustainable herding practices is happening on land planned for industrial developments: infrastructure for oil and gas, power lines, mines, and the wind industry. Whilst one governmental department says reindeer herding is unsustainable and the tundra must be protected from over-grazing, another designates the same areas as promising for minerals and energy. The Fálá district, where Sara has his herd, has summer pastures in an area of West Finnmark which entered the petroleum era with the construction of an LNG power plant in the early 2000s. The oil and gas in question, though mostly located offshore, also comes with onshore infrastructure whose construction causes increased activity and piecemeal loss of land for reindeer herding and other nature-based practices and livelihoods (Dale 2018, Vistnes et al. 2009). Gradual encroachments accumulate to make reindeer herding more difficult, and also changes the conditions for other types of traditional use of the region. This responsibility is fragmented into a thousand pieces in impact assessments, regulatory responsibilities split between state, regional and municipal authorities, as well as the industry itself. With more wind turbines and mining projects on the drawing board, the Sámi president has warned against 'green colonialism', where the Sámi are pushed away from their rights to land and to practising their culture in the name of green energy (Normann 2021). And as part of greening the Norwegian petroleum sector, Norwegian politicians want the platforms to be powered from land, by renewable Norwegian energy. Currently the power grid in Finnmark is not strong enough to power multiple offshore projects. A new 420 kW power line, currently under construction, and a decision is pending on whether it will be stretched to Hammerfest and Kvaløya to power the petroleum projects currently in operation and prepare the ground for future projects.

All these potentials and actual developments matter for future-making: with the Norwegian Arctic at the cusp of a new wave of development, centred around energy projects and mines to provide for the needed transition to a green, low-carbon future, the question of Sámi rights and land-use is once again an issue of contention and debate. Their use of the outfields (*utmark* in Norwegian, *meachhi* in Northern Sámi) are often irreconcilable with the building of infrastructures such as power lines, harbour areas, and wind farms, all of which lay claim to areas already in use by a minority whose rights and needs are too easily brushed aside as 'interests' that are less important than those of majority society. Neologisms like coexistence make it seem that the grazing land and the industrial area can be one and the same, deepening the alienation on which resource extraction depends (cf. Tsing 2015). During a debate at Tromsø Museum in January 2017, discussing the role of art in Sámi society, a Sámi politician for Norgga Sámiid Riikasearvi/The Norwegian Sámi Association, Sandra Márjá West, asked who owns the truth, when oil is supposedly green and the reindeer herders are mistreating the environment. Sara's case, then, is connected to a much larger process of neo-colonialism, or of a colonialism that never ended, but only took other forms. As *Pile o'Sápmi* puts it; *there is no post-colonial*. There can be no post-colonial, when neither past nor present is acknowledged and the structures of coloniality remain.

In recent years many scholars, indigenous and non-indigenous, have grappled with this, enunciating how other futures are nested within the conflicts over energy and industry. Àslat Holmberg's research on the fishing practices and human–salmon connections in his home region (Holmberg 2018) is one such example, where Sámi traditional knowledge practices have been outlawed by a new border agreement between Norway and Finland. The active disobedience of these laws by locals and active allies is illustrative of how futures are practised in spite of, not through, the state regulation. More than abstract questions of land rights, these struggles are about the right to self-governance and lifeworlds in a region that is still ruled through paternalistic policy-making and colonial mindsets. Though official Norwegianization policies ended decades ago, there are still several situations where Sámi are ignored, discriminated against, or disbelieved in processes of law-making or infrastructure projects, from border agreements to salmon management and industrial wind parks (Holmberg 2018, Otte, Rønningen, and Moe 2018).

What is there to do in the face of these structures? The anthropologist Georg Henriksen once claimed that 'the question of Sámi land rights in Norway appears to many Norwegians, including many Norwegian anthropologists, to be so difficult that they refrain from engaging themselves in the issue' (Henriksen 2003: 123), and this seems to continue to be the case. As a Norwegian citizen myself, I found it endlessly frustrating to continue facing the lack of knowledge about Sámi people within Norwegian majority society. For each phone call I made, each time I explained again to a journalist what the issue at hand was, or each time I encountered in everyday conversations the notion that 'there are too many reindeer, aren't there?', I felt the invisible wall of Norwegian racism throwing my explanation back at me, my words failing to penetrate the preconceptions that most Norwegians have from their lack of knowledge and scarce teaching on the less glorious parts of Norwegian nation-making. The Norwegianization policies towards the indigenous Sámi officially lasted from the 1850s until the late 1960s, but these structures continue to shape lives and landscapes in Sápmi today. Máret Ánne's artwork reconnects the abstract industrial expansions to the historical dimension and the consequences both past and present have on those who depend on the land, human and non-human. Each iteration of the artwork, whether in the form of an exhibition, a pop-up event, a debate, or a piece in the daily press, is an assertion that these are not questions of past injustice, but current and unfolding injustices nested within the Norwegian political system. The artwork's structure, and its flexibility, facilitate continued conversations and action for this system to be otherwise.

Unfinished energy futures and imagination: drilling wells, energy speculation and geology in the Chilean Andes

Martín Fonck

The below image shows the head of a geothermal drilling well in the El Tatio' geyser field in the Atacama Desert in Northern Chile. This site is the largest geyser field in the southern hemisphere, located at 4,200 m above sea level in the Chilean Andes. El Tatio is located in indigenous Atacameño territories and currently is under the administration of the community of Toconce and Caspana, through a tourist concession given by the state (Decreto Exento N° 667). This place, due to the geyser spectacle, is one of the biggest tourist attractions in Chile. However, during the twentieth and twenty-first centuries, this charismatic site has been imaged, explored, and abandoned as a potential place for developing geothermal energy. Technologies of perforation to explore underground resources have materialised 'energy speculation' and continued the extractivist logics as a colonial practice in this region. Following geothermal exploration, I want to discuss in this piece the practices of speculation, analysing energy futures imaginations. Starting with geothermal exploration for electric energy production, I show how the narratives and regulation to use this energy as a sustainable source in the Chilean Andes becomes a strategy for water speculation. Challenging the novelty of geothermal explorations in this place, and taking a historical perspective, this energy future reproduces a colonial and extractive logic. However, by analysing scientific description it is possible to follow how this same place, due to its specific geological characteristics, has resisted becoming a source of electric energy. This resistance produces a particular sense of futurity, suggesting the unfinished nature of energy futures as an alternative description.

Fig. 5.4: A geothermal well and a geyser cone. Photograph by Martín Fonck, 2019.

During the first decade of the twenty-first century, geothermal energy futures and the exploration of underground water reservoirs as a source of electricity was promoted as a potentially sustainable alternative to fossil fuels in Chile. In 2000 Congress regulated the exploration and production of geothermal energy enacting the geothermal law (Ley N°19.657). Conflicts arise because article 27 of the geothermal law states that a geothermal concession has rights over the water that may potentially emerge during the exploration process. This space of ambiguity about underground water property turned into an opportunity and an extended interest in obtaining this concession and fruitful context for speculation regarding underground water rights. These regulations and this underground exploration are part of a broader historical context of extraction produced by the consequences of industrial mining companies in northern Chile, under an extractive economic model and the consolidation of the Chilean state since the middle of the nineteenth century (Mendez, Prieto, and Godoy 2020). This logic is reproduced by the exploration of new frontiers for extraction and potential future sources of energy and resources for energy transition. For instance, in this area extractive industries are showing increasing interest in extracting lithium to implement global energy transition and the production of batteries:

> Half of the earth's identified lithium deposits are found in South America's arid, otherworldly landscape of high-altitude salares and an increase in demand would be environmentally devastating for the region (Bonelli and Dorador 2021: 3–4).

A relevant aspect in the implementation of this model has been the regulation of underground resources such as water. In this regard, an iconic example in the instauration of free-market economic policies during the Chilean military dictatorship (1973–1989) was the creation of the Water Code in 1981. An extensive literature has analysed the Chilean case as an example of the instauration of private property rights, reducing the state's role (Bauer 1998, Prieto and Bauer 2012). Water rights of use are given perpetuity to private owners, which is a particularity of the Chilean legal system. Consequently, the state has limited power to regulate them, giving scope for private market speculation regarding water rights. Going back to geothermal energy explorations, at the beginning of the twenty-first century this renewable energy alternative to fossil fuels became a potential market of exploration. The potential futures of geothermal energy and law regulations allowed access to underground water rights. However, geothermal futures have been a source of conflict in northern Chile regarding underground water potentials as a source of electricity. Geothermal resources are located in Atacameño territories, where water is sacred and mountains are considered a source of water, life, and the protection of ancestors (Carrasco 2020).

Geothermal explorations are not new in El Tatio, and didn't start only as a solution to climate change and green transition. Under a broader historical context and to promote industrial mining, the first geothermal exploration for electricity production was registered in 1923 (Tocchi 1923). The agenda of this exploration was characterised by bringing scientific and technical knowledge from Larderello (a site located in Italian Tuscany, where the first geothermal electric plant in the world was located) to this place. During the exploration, they carried out three drilling wells, establishing the first approximation to the geothermal futures of the area (Morata 2014). Engineers' imaginations and dreams were motivated by the will to provide electricity to the province's industries. While searching for this resource, they imagined a future electric landscape. However, these explorations were left behind due to a lack of funding and technical problems.

As stated in Italian engineers' reports, the main challenge of this exploration was framed as 'the problem of intermittency'. The eruption of geysers in El Tatio varies according to the time of day. During the coldest hours of the day, the steam begins to emerge at night, but with the sun's first light, it decreases and sometimes even stops. This feature was continuously compared with the Italian geothermal site of Larderello, where geysers constantly emerge during the day. The landscape of El Tatio was read through the lens of Italian Tuscany. The geysers were described as an intermittent phenomenon, which expresses itself in variable and unpredictable ways, destabilising the imaginaries of constancy brought from the Larderello landscape. The variable presence of the steam at the surface was the focus of this exploration. The Italian engineers imagined the future presence of infrastructure to transform this energy into electric energy. A source contained in the underground, which can be reached by drilling wells. However, the intermittency of the phenomena caused them distress and anxiety due to the possibility of failure in their endeavour of producing electric energy.

The geologist Juan Brüggen (1940) showed a different perspective about this envisaged energy source included in a report commissioned by the state where he described these first explorations and perforation. Brüggen, one of the founders of the discipline of geology at the University of Chile, describes the intermittency of the geysers' steam as a geological incognita. By studying the landscape, he elaborated hypotheses and measurements to explain the absence of steam, transforming it into a scientific challenge. The winds, water temperature, and their glacial origins were selected as possible factors to explain what happens inside the earth. Although his geological report was used at various times to argue and promote the potential of these sites to produce electric energy, in the same document he is sceptical towards the possibility of producing electricity, due to the intermittent geological characteristic of the place.

Nowadays, the traces of these geothermal energy futures are visually present, such as the one with which I started this piece. Tourists often take photos of these old infrastructures that have become part of the landscape's visual attractions. The way of bringing underground water into the surface through drilling wells in the place has not worked. Although the traces suggest the materialisation of an energy future, at the same time they also embody how this logic has not achieved its goal, moving the economic interest and speculation to other areas for the production of electricity (mainly solar and wind energy). Although the recent socio-environmental conflict regarding the visual impact of geothermal exploration has played an important role (Fonck 2021), the unfinished nature of this energy potential produces a different sense of 'futurity' and imagination. The intermittency of the phenomena and their relational ecologies, described in the geological reports, suggests the elusive trajectories of underground water, challenging the energy future produced by perforations technologies, and their imaginaries of constancy and electricity.

Tracking frackers: the messiness of prospecting for futures in the Kalahari

Pierre du Plessis

Tracking the emergence of resource frontiers can be tricky business. It is not always clear when the process begins and the signs are often specifically located. Such processes are a form of 'slow violence' (Nixon 2011) that do not erupt with immediate effect, but emerge slowly, cumulatively, and often out of sight with distant repercussions. They are not always visible to broader publics until it is too late. The processes involve a great deal of speculation, are often bumbling in nature, and can be quite messy in practice. It takes a lot of work to instantiate landscapes as 'standing reserves'. And indeed, making landscapes into 'standing reserves' is a messy business.

In Botswana's Kgalagadi, which means 'land of thirst' – more widely known as the Kalahari Desert – a different kind of tracking drew my attention to the bumbling nature of natural gas prospecting precisely because of its messiness. I was studying tracking as a skilled practice of noticing the presence of multispecies actors – animals, but also insects, plants, and fungi – by the tracks and traces that they make in landscapes, learning to interpret behaviours, and how to anticipate movements based on those tracks and traces (Du Plessis

2018). Tracking is a skill that is well known among San-speaking people in the Kgalagadi, and that has been utilised as part of a set of knowledges that support-ed hunting and gathering activities in the past. Despite restrictions that have lim-ited their hunting and gathering opportunities, people continue to track today; they track as means of employment in tourism and wildlife conservation proj-ects, but also in their daily lives as a way of knowing, relating to, interpreting the landscapes with which they live and the many doings of humans and non-humans that share these environs. Tracking, as I was coming to understand it, is a mode of attunement towards the movements of landscapes and the relations with which landscapes hang together (Du Plessis 2022). It is not limited to the pursuit of hunted animals moving across a backgrounded landscape. It is attune-ment through tracking that, today, helps to tell the stories of new encroachments and developments that threaten Kalahari landscapes and, more often than not, threaten the livelihoods of the people who live in remote regions of this desert.

Tracking attends to landscape patterns, and when those patterns change or are interrupted it is as if one is compelled to notice them. In a 'thirst land' that is increasingly encroached upon, habitats fragmented, and people dispossessed, tracking is also a mode of noticing socio-material transformations that have emerged, in large part, with histories of colonialism that resonate into the pre-sent. Histories that have treated people and landscapes as resource reserves. The processes are ongoing, even if they have become so normalised as to have the effect of being 'out of sight'. Tracking foregrounds these material traces to bring them into view, resisting the static depictions of territorializing maps that facilitate land grabs. Tracking reminds us, if only we take the time to notice, that 'there is no postcolonial', by drawing attention to landscape patterns and the ways that they are interrupted when made into resources.

It was while still learning to track, and as I was beginning to grasp the extent to which it is a critical skill for understanding more-than-human worlds and re-lations, that a new set of movements and interruptions arrived to the region where I was conducting my field research: those of natural gas prospectors. This area, known officially as the KD2 Wildlife Management Area (WMA), is lo-cated in the 'Schwelle', a region characterised by a lack of surface water, relative-ly open savannah grasslands, deep Kalahari sands, and a dense concentration of salt pans – mineralised depressions that that only hold shallow water briefly after rains – and the tall, stabilised, dunes associated with the pans, called lu-nettes. It is a critical ecological habitat for Kalahari wildlife, and the pans, espe-cially, play an important role for the desert wildlife in this thirstland, providing occasional water, mineral licks, and supporting nutrient rich vegetation.

The KD2 WMA is a remote, but extremely significant link in a wildlife corri-dor that stretches some 700 kilometre between the Kgalagadi Transfrontier Park

(KTP) and the Central Kalahari Game Reserve (CKGR), perhaps the world's longest, uninterrupted, wildlife dispersal area. The corridor, however, is steadily narrowing, encroached upon by the growth and development of villages, but especially the steady march of cattle and the growing cattle industry that have come to feed on Kalahari grasslands more and more. Indeed, much of my own research, and the efforts of trackers with whom I work, has been dedicated to drawing attention to this area in order to slow the closure of the wildlife corridor (cf. Keeping et al. 2019, Du Plessis 2010). So, when signs of heavy machinery began to appear, leaving deep scars in the sand tracks and pummelling vegetation as it made its way through the bush, there was cause for concern that this already vulnerable, but important, wildlife area might be compromised by a new development project. What occurs out of view can have long term, lingering effects.

This contribution follows the first stages of a prospecting venture that explored parts of the Kalahari for natural gas, but ultimately failed in large part due to the resistance offered by a lively desert landscape. This is not a success story, but a story of failures and environmental destruction. It tracks the early stages of extractivist projects long before they succeed at reaching their extractivist goals, and highlights some of the early effects of enacting resource frontiers. It offers an example that shows how landscapes are lively actors that prospectors and extractivist industries seek to overcome, whose speculations alone garner huge investments from those looking to secure their capitalist futures. It also offers a glimpse into how failed prospecting projects serve as signs of the emergence of larger extraction projects in the future, while having immediate and long-term social and ecological effects that transform relational landscapes into territories. That is, by the time prospecting occurs, which often happens with little fanfare and out of sight, the frameworks for extraction projects are often already well on their way to being realised. The social-environmental consequences, in turn, appear to be treated as mere collateral and expendable debris, even if companies and states make a distinction between prospecting and actual extraction to defend their early efforts. Erasures do not happen suddenly.

Early reports

In late 2015, a report emerged claiming that fracking was occurring in some of Botswana's protected areas, specifically claiming that these activities were occurring in the KTP, but also potentially the CKGR (Barbee 2015). This was not the first time that such claims had been made, though they were largely rebutted by the Botswana government (Barbee, Dutschke, and Smith 2013). The report, published in *The Guardian*, was authored by Jeffrey Barbee, a filmmaker and environmental activist who had already produced a film in 2013 entitled 'The High Cost of Cheap Gas', (https://www.opensocietyfoundations.org/voices/high-cost-cheap-gas-southern-africa) critiquing the emergence of natural gas extraction industries emerging in South Africa and Botswana. In response, in 2013 the Botswana government denied these claims, with one official saying:

> There is currently no fracking in the CKGR or anywhere else in the country. Coal bed methane is being prospected in the country, though there are no current commercial operations. Still not clear whether commercially viable or not, though a number of companies have shown interest...Coal bed methane prospecting in the country is being carried out by drilling, not fracking. Environmental impact assessments and management plans are a legal requirement for all mining in our country. Any future proposal for a licence for commercial fracking in this country would certainly be subject to vigorous debate, whether in a game reserve or anywhere else. (Staff Reporter 2013)

This statement, however, was less a denial than a contestation of the definition of fracking, and the promise of 'vigorous debate' was ultimately unrealized. The state claimed that because mining companies were not yet extracting gas via hydraulic fracturing but opening exploration well holes by drilling, that these practices did not amount to fracking. According to more recent statements made by another company, Tlou Energy, that has prospecting rights in another region of Botswana to the west of the CKGR, it produced its first gas as early as 2014, while another company, Kalahari Energy, claimed to have prospecting rights as early as 2012. The link between prospecting and the eventual move towards extraction and production are clear.

The effects of simply prospecting, however, as Barbee's film showed, are also significant and have their own environmental consequences, not least of which offers the spectre of gas fields in the future. By the time Barbee's 2015 article was published, enough evidence had emerged about the extent of natural gas exploration that the state amended its response to say only prospecting was occurring and it was not occurring in the parks, occluding the fact that the wildlife areas between the parks are semi-protected and communities have lease rights to these lands. Indeed, details emerged that natural gas prospecting licences had been sold for huge swaths of land, including to a company called Nodding Donkey that was later rebranded as Karoo Energy. The company boasted a map on its website showing the extent of its prospecting licences to cover nearly the entirety of the wildlife corridor between the CKGR and KTP, including KD2, as well as parts of the parks.

In all, the licences covered an area of more than 140,000 km², a massive swath of land that led the CEO of the company, which operates under the moniker of a variety of subsidiaries, to state:

> We are delighted at the awarding of these petroleum exploration licences by Botswana's Department of Mines. We are excited at the potential of the licences to host shale gas, which can play a vital role as an energy source for electricity in Botswana and the wider African region. We look forward to commencing work on the licences shortly. (Stock Market Wire 2015)

This statement, coming two months before Barbee's article, was directed towards international investors, but the Botswana public remained largely unaware of these operations.

Tracking the Prospectors

As these reports emerged, I was in the Kalahari working with a group of master trackers who were teaching me how to interpret landscapes through the practice of tracking, and our attention had shifted towards following and finding particular plants that my interlocutors gathered to eat or use for medicinal purposes. While resupplying in a small town in late December 2015, I received a phone call from a researcher working in the same area to tell me that he had come across signs of heavy machinery moving through the bush, cutting deep and wide tracks on an old sand trail towards a pan several kilometres outside a small settlement called Zutshwa in KD2, the vast, unfenced wilderness adjoining Kgalagadi Transfrontier Park. He followed the trail to its conclusion but found only an abandoned camp and a semi-sealed coring pipe sticking out of the Kalahari sand. We immediately suspected that this must have been the work of natural gas prospectors. We wondered where they had gone and when they might return, and what exactly it was that they were doing or looking for. My colleague asked a resident of Zutshwa to phone us if a drilling team returned so we could investigate. In light of these new developments, I would track the frackers.

In early January 2016 I received the call that a convoy of trucks had returned and I drove to Zutshwa as soon as I could – this time without my tracking teachers – worrying that the trucks might leave before I arrived. On my way to the settlement from a larger town called Hukuntsi, I passed several large trucks carrying equipment along the 60 kilometre, bumpy gravel road. When I arrived at the settlement, a team of contracted labourers had already begun setting up a camp across from the local clinic and Zutshwa's water tower near the centre of the village. Their trucks and machinery were parked by their camp, displaying the emblems of 'Discovery Drilling'. The drilling team had arrived. Before approaching them, I spoke to community members to ask about the activities of the drilling team. I was told that they were looking for water, which most people were happy about since they often encountered infrastructural problems with the pipes that filled their water tower. It would be great to have a more reliable source of water, several people told me. Others, however, thought that the team was wasting their time because they knew that the ancient groundwater aquifers beneath Zutshwa pan, situated next to the village, would be too salty. After a few days I was able to speak with one of the people camped with the drillers who told me that they were looking for water because they needed it to lubricate their drills when they did deep drilling to extract core samples.

The team soon realised that the groundwater they were able to locate was indeed too saline to use in their drilling machinery and refocused their efforts on moving their equipment to their primary drilling site located on the edge of another pan called Gonkhu, some 20 kilometres into the WMA from Zutshwa. The deep sand and thick bush, however, were significant obstacles as the team struggled to lug their equipment, and ultimately used a JCB (earthmover/bulldozer) to clear the way, knocking down small trees on the way to the site.

Fig. 5.5: A JCB bulldozer clearing track to drilling site.

Discovery again attempted to find water near Gonkhu pan, but was unsuccessful. Frustrated, a team went in search of an old government borehole near Peach Pan, almost 50 kilometres away from Gonkhu. This team was led by a geologist who followed an old hunting track all the way from Gonkhu to Peach, where their tracks indicated they stopped to look at the sealed borehole, then to Zonye, Towe, and Name pans, and then back to Zutshwa, a more than 80 kilometre loop through this semi-protected area.

The foreman of the drilling project, a Motswana man in his late 40s or early 50s, revealed to me that he thought that they would have to get water from the old government-drilled borehole at Peach Pan because he 'can't see any other way'. He seemed really discouraged when I asked him how they would do it, considering Peach is almost 50 kilometres away, and just shook his head, and said 'There is too much bush and sand here.' Clean water was crucial for their drilling, which they mixed with a chemical solution he called 'EZ Mix' as they extracted core samples. In the meantime, he told me, they would truck water in from the Zutshwa water tower, but they would need another, more robust source, for he was also quite sure that they would sink more coring wells, especially if the core-sampling showed positive results. At the time, the Gonkhu well was 200 metres deep, but their goal was to take core samples up to a depth of 1,000 metres.

I returned a few weeks later to find that despite the prospectors' best attempts they had made little progress in reaching their target depth of 1,000 metres because they had exhausted the requisite water supply from Zutshwa to lubricate their drill bits. They had resorted to trucking in all their water from Zutshwa. And, because the deep, porous sands meant that the coring holes did not retain much of the water, they were using 30,000 litres of water a day from nearby Zutshwa, a settlement already characterised by water-scarcity, to advance their drilling only 10 metres per day. Unexpectedly, one of the drilling contractors told me, they were not able to recycle the drilling water that remained in the coring tube as they normally would when drilling into other formations. The drilling crew also had to abandon their first coring hole when the tube became stuck in the hole and they had to start over, drilling a second well only a few metres from the first. Ultimately, if they were going to reach their target depth of 1,000 metres the drilling team would need to use as much as 3 million litres of water, as 30,000 litres only equated to 10 metres of progress.

Water is scarce in Zutshwa. The water piped into the settlement's water tower travels from more than 60 kilometres away, and is the community's primary water source. The drilling team used massive amounts of this water, injecting it back into the ground as a chemical slush in their search for gas. But it wasn't enough. By March 2016 the frustrated prospectors had packed up and

Fig. 5.6: Truck and trailer carrying coring pipes, immobilised after getting stuck trying to cross Gonkhu Pan.

moved to a different exploratory site, leaving a trail of destruction in their wake. Bulldozed tracks, an abandoned camp, and coring pipes littered the site, but it was the thirsty drill that caused the most damage and would have the longest effects. Out of sight, these prospectors were part of a broader project to create Botswana's energy futures that enacted this desert landscape as a potentially prosperous resource frontier, in spite of the anticipated environmental effects that would go unnoticed.

I was told that the prospectors left because of the lack of water needed to drill. A member of the Zutshwa community trust told me that the trust met with representatives of the gas prospecting company and refused to grant them access to any more community water. But Zutshwa's water infrastructure was already damaged and the settlement was left waterless, unable to pipe water in from Ngwatle. When I returned to visit a year later in January 2017, there was still no permanent water supply in Zutshwa and its roughly 600 residents' only recourse was to purchase drums of water from entrepreneurs in the nearest village, Hukuntsi, some 60 kilometres away. Even though prospecting occurred over just a few months and did not seem to be successful, it had the effect of exaggerating a landscape of water scarcity and transformed water into a commodity for the remote settlement of Zutshwa.

The porosity of the sand, water-scarcity, and the geomorphology of this particular area combined to make gas exploration unexpectedly challenging. The greatest obstacle to the prospectors was that the sand absorbed the water–chemical mix used to lubricate the drill at an alarming rate. While mining has been successful in other parts of the country, this region is extremely rich in salt pans and dunes, and has a very deep layer of Kalahari surface sands. These things together, along with evaporation, function to sap up surface water that arrives in the area. As the prospectors pumped water into the sand to lubricate their drills, the water was almost immediately commandeered by the desert drainage system. Together the geomorphology, water, and sand resisted the drilling, but despite this resistance the work of the prospectors had lingering effects. By the time the prospectors left, the water supply in Zutshwa had been exhausted.

Though prospecting had failed in this region, the results were more ominous elsewhere in the country. By 2021 Botswana's denials about fracking seemed to be long forgotten, and the effects of prospecting KD2 hardly appear as a footnote in the story. The Botswana government has since signed a deal with Tlou Energy to build a 10 megawatt (MW) natural gas power plant, albeit in a different region of Botswana than described here. It is part of a larger 20-year plan to add 600 MW of energy as Botswana 'looks to wean itself off imports while also possibly exporting power' (Reuters 2021). Over a nearly ten-year period, we see the shift from the spectre of CBM prospecting to the actual realisation of a fracked, natural gas power-plant. But the prospecting in KD2, out of sight and bumbling in nature, was part of a broader configuration of shifts and movements that contributed to the emergence of a full-scale embrace of natural gas extraction and energy production. The long-term consequences in one remote region where prospecting failed appear as mere collateral. Indeed, making landscapes into 'standing reserves' is messy business. As Fonk suggests, 'energy speculation' continues extractive logics as a colonial practice.

Data as care: the charged issue of data and energy infrastructure in a Capetonian township

Karen Waltorp

> A phone without data is no phone
> (Ricardo, June 2020)

Very far from Silicon Valley, in the South African township of Manenberg on the wind-swept Cape Flats outside Cape Town, access to both wifi-signal and energy to charge your phone battery, and not least ability to pay for these costs, sheds light on how energy futures look very different depending on where you look *from*. For this precarious assemblage to hold together, coal and wind and the extraction of earth's resources are required. It becomes apparent how people *make up* part of the infrastructure that allows the smartphone to work. To provide data needed to be part of this new technologically mediated world is a way to show care. People *are* infrastructure (Simone 2004) conceived in the broad sense of the word: the smartphone, when thought of as a node in the infrastructure needed to enable it to work, only functions when people help out each other financially with data and with charging each other's devices when there is electricity available. The phone is a relational *and* relating device (Waltorp 2020), and the work that goes into making the 'phone-assemblage' work, is part of this.

Infrastructural disparities become more visible in times of crisis; such as the energy crisis in 2007, the water crisis in 2017–2018, and the current Covid-19 pandemic, which has seen the country on strict recurring lock-downs. In this current situation, having a smartphone – charged and topped up with data – is necessary *and* difficult to attain for Manenberg residents. The unequal access to 'log on' and tap into the technological communications infrastructure and related imaginary of the modern citizen-consumer relates to data, money, and energy – and their interlocked histories. Different imaginings of energy futures emerge from situatedness over time (see also Pollio and Cirolia in Ch. 3 for a related discussion from the vantage point of South Africa's data centre industry). I have carried out long-term fieldwork and filmmaking in this township area of around 80.000 people since 2005–2006. I have revisited on shorter field trips seven times, the latest in December 2018. I have followed media use, in situ, over the phone, and in social media, tracing the radical development brought by the advent of smartphones. The communication in-between field trips occur mostly in the platforms WhatsApp, Instagram, and Facebook, configuring *fieldwork as interface* (see Waltorp 2018). In Manenberg, smartphones – and the infrastructure they connect to – are related in intricate ways to a local (moral) economy of sharing resources, thereby showing whom you are related to, indebted to, and care about. With calls, texts, chatting, liking and commenting – as well as data bought, shared or transferred to someone's phone – you show that you wish to invest in the relation. You show that you care.

Fig. 5.7: Collage from the ethnographic film *Manenberg* (Waltorp and Vium 2010) showing architects planning the townships on the Cape Flats outside Cape Town, South Africa according to the logics of apartheid's 'separate development' and classification of 'races' into 'Black African', 'Coloured', 'Indian', and 'White'.

Smartphones in Manenberg

Smartphones are cellular phones with a mobile operating system (IOS, Android, Linux-based etc.). They combine a computer operating system with the features of a mobile phone: calling, text messaging, calendar, media player, video games, GPS navigation, alarm clock, and last but not least apps (third-party software components), of which the social media platforms are important for both networking for conducting business or finding a job, and for socialising and entertainment. Today about 20 to 22 million people in South Africa use a smartphone, which accounts for about one-third of the country's population. The number of mobile connections is much higher, though, with more than 90 million (feature

5 Imagining energy futures beyond colonial continuation — **199**

phones are still widely used). The number of South African smartphone users is forecast to grow by more than five million until 2023.

Smartphones are used in a myriad of ways in Manenberg as in many other places in today's technologically-mediated world. They are ubiquitous, figuring as central networking devices to people inside and outside the township and in broader transnational networks, for business as for pleasure. The smartphones shape, register, and impact what is communicated through them; they are used for accessing information and for communication, as a tool for protection, gossiping, alliance-making, flirting, bullying, or trafficking drugs; for reading the Qur'an, or sharing Bible quotes; for geo-tagging (subversively), paying bills, raising money or mobilising for causes locally; for communicating with municipal offices and social services, and for entertainment not least, and 'flashing' one's economic and technological capital. The touch screen and related language with *emoticons*, *gifs*, and *memes*, add to the popularity of the smartphone in Manenberg: you can be fun, 'roast' someone on social media; perhaps a 'baby daddy' who forgets his responsibilities, or a friend who let you down. You can send layered messages by combining images, text, and emoticons, or by sharing memes.

In a first for the continent, Facebook opened an office in Johannesburg in 2015 (more than 80 per cent access Facebook from their smartphones). The narrative is the well-rehearsed oft-iterated mantra about *helping through connecting people*. Nicola Mendelsohn, Facebook's vice-president of Europe, the Middle East, and Africa (EMEA) stated firmly on that occasion: 'Mobile is not a trend. It's the fastest development in communications we've ever seen. This couldn't be truer in Africa – where so many people are mobile-only.'

There have been major shifts since 2005 when I first conducted fieldwork in the area and people used cellular phones (now called a dumbphone or feature phone) with Mxit downloaded on almost every phone, allowing for very cheap texting, games etc., and the proliferate exchange of 'please call me' texts, and missed calls, which indicated that you want to be in contact with someone, and *they* ought to call you and incur the expense of the data. Each 'please call me' text message is a risky claim on the person and a hope invested in the relation. Furthermore, across the grey zone of illicit activities that exist in marginal spaces where people must make do, the phone adds to this and makes porous, usual and assumed walls and borders between things. Money passes more or less freely into the prison from the outside, illustrated by a visit to a friend and informant in Pollsmoor prison outside of Cape Town. Asked if I had any money on me, and my affirmative answer, he asked me to please give it to him. Transferring money to a hidden phone as inserted data, works as currency inside the walls. Although data cannot be transferred into cash officially, they *are* a currency, both in symbolic and literal terms, as alluded to here. These 'awkward entanglements' in the area between prison and township, the law-abiding family members and those who commit crime (Waltorp and Jensen 2019), are born from the long history of disenfranchisement, colonialism, and apartheid rule. You can support friends and family living, working or travelling in South Africa by sending the same 'airtime top-up' (data) directly to their mobile phone. For this infrastructure to work in marginal places, you need people as infrastructure – digital–material infrastructure – and the electricity to power this assemblage.

The smartphone plays into both activist engagement and mobilising in the area, as well as the need for escapism. What people across these various uses share is that they need data to be part of those different conversations; they need energy to charge the phone battery and an infrastructure that allows for the smartphone to be a connecting and relational device, audiencing and communicating within and beyond the confines of the township. The pleas from a child to a parent start early: 'Please, mommy, buy me data', when the favourite cartoon stops in the middle, or a game is exciting and abruptly cut off. Once you have depleted the data resources, your internet session will be paused and a message will be displayed indicating this. Generally, you can buy data in voucher form from any corner shop, supermarket, or cellular network store in South Africa. When you use your data to go online, you pay out-of-bundle rates, which are much higher than in-bundle rates. Data bundles give you a set amount of data for a fixed price, but it means you need to have a larger amount of money to pay up front, which is often not the case for people in the area. An often-heard lament is that buying more data means using 'bread money'. This is telling in terms of how difficult it is, when this enlargement of our experiential world which the smartphone affords, disappears. Even more so in marginalised areas where only a third of households own a car, public transport is expensive and very time-consuming, and few have a stationary computer or a laptop. The smartphone is what can connect you – for business and for pleasure – with the outside world. Even if you succeed in getting the money to go online, or borrow a relative's phone when things are looking down, if the phone's battery is flat and cannot be charged, it is then useless. To paraphrase Ricardo, whose quote opened this ethnographic case; *A phone without data is not a phone.* But where does the electricity come from to charge the battery? And where does the signal come from?

Without the electricity, the mobile networks were under strain in early 2019. In periods of load-shedding – scheduled electricity blackouts by the national electricity provider Eskom –the signal might drop on the phone too, leaving you unable to make calls, send and receive messages, and load anything, even if you managed to charge your phone before the power cut, or in alternative ways. As Susan Leigh Star observes, infrastructure is typically 'invisible' until it breaks down (1999: 382). Even if a person charged their phone before scheduled load-shedding, they might find there was no reception either way. For the cell towers to emit a signal, it requires a power source: Eskom. Even if towers use batteries as a backup, these have limited power and will eventually be depleted.

In Manenberg access to data is expensive, bad, and unequal. Around the year 2000, Western Europe and North America were enjoying the benefits of well-developed fibre infrastructure. In 2015, South African broadband costs were up to ten times higher than the UK and local speeds were five times slower. South Africa left the Commonwealth in 1961 and so missed out on the COMPAC Commonwealth telephone cable. Instead, the South Atlantic Cable Company was formed in South Africa by the Industrial Development Corporation of South Africa and American Cable and Radio. The SAT-1 laid in 1968 was replaced by SAT-2 in 1993, and later the SAT-3 cable was operational from 2001. One single undersea cable serviced South Africa (operated by Telkom) until in the last decade, when this monopoly was broken with the introduction of Eassy and Seacom.

The current price of fibre in South Africa is likely the most expensive it will ever be, due to high outlay and maintenance costs associated with light-transmitting fibre optic cables. The first locations to receive fibre coverage locally were more affluent areas, where people could afford it, and infrastructure companies could more easily make back their investment, and only a small section of the population have access to it; the relatively well-served urban and suburban areas that make up only 7 per cent of South Africa's geographic area; while the remaining 93 per cent is rural, and generally poor, and not served by telecoms infrastructure (yet). The South African government intends to replace all copper cables with fibre optic cables, and secure access to this infrastructure for all by 2030, which commentators view as optimistic to say the least. Facebook is underway with the planned *2Africa* undersea fibre optic cable network, which is to drive down costs and have people sign up (and for Facebook to access their data).

As mentioned above, charging your phone can at times be quite a challenge, as load-shedding has become a reality for South Africans, severely impacting daily routines. Africa is the world's least electrified continent and its electrification rate is growing more slowly than anywhere else. As in several other African countries, South Africa's problems are rooted in being unable to expand electricity infrastructure quickly enough to cope with population growth and demand.

Electricity in South Africa was publicly used for the first time with the opening of the electric telegraph line between Cape Town and Simon's Town on 25

Fig. 5.8: Aerial photograph of the township Manenberg, from the ethnographic film *Manenberg* (Waltorp and Vium 2010).

April 1860 to support communication infrastructure, and for the benefit of shipping and commerce. Today, the government-owned national power utility Eskom dominates the country's electricity sector. With 27 operational power plants generating over 95 per cent of the country's electricity and over 40 per cent of all electricity on the African continent, Eskom is one of the ten largest power utilities in the world. However, the South African government has been unable to keep Eskom's generating capacity up with economic and population growth. Lack of investment by the government has resulted in energy shortage and led to an energy crisis in late 2007. This has forced Eskom to implement load-shedding in specific areas of the country at certain times to reduce pressure on the national grid, and to initiate an ambitious program to increase energy production. As part of this, the Cape Region, where Manenberg is situated, wants to lead the drive for a future where coal plays a lesser role, as wind-power installations with an additional 3.3 GW will be added to the region's energy capacity by 2024. As in the other ethnographic cases in this chapter, green energy captures the imagination and 'hides' its costs.

Imagining energy futures from Manenberg

In the context of Manenberg township in South Africa colonial/apartheid infrastructural arrangements continue in terms of how energy, as well as network (connection), are distributed. Those located in disadvantaged areas due to apartheid-era policies are similarly disadvantaged today in the current energy situation: They need to be connected as much as the next person. In fact, even more so, as casual labourers lose out without a phone, and those who wish to network and make a future for themselves need the phone to connect, and for that connection to be reliable not least. Without a phone – and a phone without data is not a phone to reiterate Ricardo's words – it is as if you are being decoupled from the outside world and the local, intimate world alike. You fail to live up to, and maintain, your relational work locally, and the expectations on you to provide for children, spouses, lovers, or friends. Awkward entanglements and crucial connectivity in the (moral) economy in Manenberg play together with ways of showing care by 'investing' in data; spending them on each other, or giving them to each other to spend.

In local conversations about the recent energy crisis, water crisis, and the current Covid-19 pandemic, you often hear a reference to how it's always 'Manenberg se mense' (people from/in Manenberg) who suffer, are disconnected, and experience lack when resources are scarce. Manenberg residents imagine their energy futures in various ways; they protest when the water crisis hits them first (Robins 2019), they hope for cheaper prices on data, and for cheaper electricity overall. Indeed, the electricity bill is a fraught subject in many households. Local practices and imaginings in relation to smartphone use and data use are far removed from the emerging (imagined) energy situation of President Ramaphosa's promise in 2020 that he would break Eskom's monopoly, and move away from coal as the main energy source and invest in windmills. Wind energy as a solution for the future and present lack of energy in South Africa's Western Cape region seems to bypass and suppress other urgent conversations. In the meantime, local residents share tips on how to charge your phone when there is load-shedding, and not least develop tactics to make the little money they have for data last the longest, by for example make sure to send each other 'please call me texts' (albeit many will state that this is a thing of the past as no one wish to seem 'cheap'). People will share images and other 'heavy' items via WhatsApp and not in MSN, and if possible when connected to wifi at work for example. The standard advice in relation to load-shedding, which you will find on the network's web pages read: When load-shedding occurs and you do not have enough battery life, you can always charge your device in the car (presupposes that you have a car); make sure you always carry a power bank or two as they will come in handy (if you can afford it); make sure to keep your laptop charged (if you have one), so that when the lights go out you can charge your mobile devices on it.

These practices do not apply to Manenberg mense, and are very out of sync with what is possible locally. In taking a look at energy futures from Manenberg, I wish to point out how aspects of the unequal structures that go into the development, spread, and consumption of specific technologies, here the smartphone, are obscured by the names we know them by. The telephone is *smart*, and the data called *air-time* we buy to carry our voices, words and images to each other, is not in and of air. It is the ground we should look to: it is bound up with resource extraction, in South Africa it is coal, and it brings a host of problems to the environment and people living in it. It takes some untangling to perceive the obscuring poetics of the digital–material entities and services, such as the cultural fantasy of the cloud (Hu 2015): as stated in Ch. 4, there is no cloud, there are material infrastructures in the form of bunkers and fibre optic cables laid in the ground and under the sea. The previous infrastructures of railways and telephone poles become apparent, and oftentimes run along the same lines. These (colonial) infrastructures are material and come with related imaginaries. Each in its own time has carried forward the story of technological development as inherently good and desirable, and the specific technological future inevitable, with no further mention of who benefits (see also Benjamin 2019, Star 1999). To imagine energy futures differently implies situating oneself far from Silicon Valley, and look at the continued material structures, and how they relate to what *can* and *cannot* be imagined by whom.

How to address the challenge?

All four ethnographic examples discussed above use various strategies to 'make visible' what is hidden or out of sight and thus little discussed, or discussed in ways that mis-recognizes some aspects of the struggles and frictions around energy. The cases seek to make specific contested energy futures into 'issues of concern' (Latour 2004, Marres 2005). We bring these aspects into the (academic) public's view, and thus question the (non)power of the anthropologist to communicate and make things public in ways that go beyond narrow academic circles and readership. Imagery and visuality is a modality that travels so much faster than writing and reaches other audiences (Waltorp 2021). Simultaneously, imagery and visuality travel in different ways not easily controlled, thus making questions of intervention, positionality, and ethics as non-Indigenous anthropologists even more pertinent.

As in the European Association for Social Anthropologists' Future Anthropologies Network manifesto (FAN 2014 https://www.easaonline.org/networks/fan/), we see the role of critical ethnographers as confronting and intervening in the contested futures of our different field sites and staying with the complexities and differences across scale, places and power relations. This can be read as a form of collaboration across difference, with varying understandings of what that means for different parties involved: We find some resonance with Kim Fortun's notion of 'enunciatory communities' and Hall's discussion of 'articulations' joining various positions of enunciation (Hall, Morley, and Chen 1996). Such enunciatory communities emerge in response to particular situations whilst cross-cutting them as heterogeneous, complex phenomena (Fortun 2001). In different ways, all our examples show a collaborative commitment to a future that may still be otherwise. As Dale shows in her ethnographic case above, artworks can carve out an alternative arena for future-making and struggles over energy and what justice and fairness means in a region and country which is planning for a future without coming to terms with its colonial past and present. The Pile o 'Sápmi-project brought together Sámi, non-Sámi, artists, anthropologists, photographers, community educators, curators, freelance journalists and documentary makers with different agendas, commingled into action during weeks of manifesting a curtain of reindeer skulls outside the Norwegian Parliament. They also stretch across and beyond to all the other places where the work has manifested or echoed. It has been part of reshaping organisations like the Office for Contemporary Art in Norway, who have made themselves an active ally in Sámi struggles, and the National Art Gallery, which bought the artwork to exhibit to the Norwegian and international public. Cumulatively, it has created an awareness of Sara's case that goes far beyond most contemporary struggles in Sápmi, whilst also connecting the dots of all the different regulations and extractivisms, from power lines to mines, cabins, and petroleum. The pile of bloody reindeer skulls is both a warning of a crime the government is in the act of committing, and an intervention to shape the future otherwise. Art mobilisation, then, becomes a way of working through and building action that makes different futures possible, and these arenas of enunciatory collaboration a possible place for a more liberatory anthropology to start (Harrison 2011).

The case presented by Fonck challenges the narratives of geothermal energy futures as a novel source of electrical power. Fonck does this by focusing on the iconic site El Tatio in the Chilean Andes, where geothermal energy has been imagined at different historical moments. This case highlights the role of geological entities and their description in asking how energy futures are imagined and the colonial logics they reproduce. Starting by presenting a general overview of strategies and regulations of geothermal energy that, under the narrative of facilitating the use of new energy sources, enable new frontiers for speculation. These strategies are linked to a broader context and history of extraction. In addition, the first exploration that imagined the underground of this place as a source of electricity observed the site through European lenses and references, comparing the geological characteristics of this Andes site with the Italian region of Larderello. However, by examining the geological descriptions, it is also possible to find a different perspective, namely the resistance of the geological phenomenon to becoming a source of electrical energy.

Du Plessis shows how the processes through which remote landscapes are transformed into resource frontiers involve a great variety of often contradictory ideological framings, practices, and enactments. Not least of these are the simultaneous speculative practices that treat landscapes both as empty, homogenous, and static and as full of untapped potential, diverse, and lively in order to enact them as 'standing reserves' before resources are located. This work usually occurs out of sight of the public, but with dramatic effect, with the specific joint aims of striking it rich and securing energy futures through the potential extraction of CBM gas, even as the ventures fail to produce. The notion of reserves, with the attached implications of preservation for exploitation, initiate temporal disjunctures that enact imaginaries of capitalist resource futures, which persists in the seemingly opposite liberal spectrums of conservation and extractivist discourses. In Botswana, where mining and wildlife tourism are the country's two most profitable sectors (along with the beef export industry), the worlds of wildlife and energy 'reserves' come to clash, competing for geographic and ideological space in Kalahari and its landscapes. Both inhibit and affect San livelihoods and environments in different ways, but often play out in the form of restrictions limiting access to their own lands for these communities.

The emerging new energy landscape of South Africa, and the Cape Region in particular, which Waltorp focuses on, emerges in/from a world already structured socio-culturally, economically, and politically – in racialized ways. Ruha Benjamin (2019) invites us to think about the way that race and technology shape one another and how the ethical and social impact of technology, which is only half of the story, is always already about social norms, values, and structures existing prior to any technological developments. Most people are forced to live inside someone else's imagination; elite fantasies about efficiency, profit, and social control, to which Yolande Strengers (2013) also points with her coinage of the term 'resource man', focusing on smart technology and the citizen imagined in this scenario – an imaginary in which is embedded both ableist, sexist, and racist attitudes implicitly. As underscored in this book's Introduction, to better invest in and develop energy infrastructure we need to keep people in mind. Recent examples of smart city failures and growing demand for energy have demonstrated that new thinking is needed on how infrastructure is designed and built; digital infrastructure investment should be fully engaged with people in their diversity, complexity, and everyday needs. Overtly utopian or dystopian scenarios prevent us from exploring on the ground realities connected to energy futures, which can and should be considered to be able to intervene in meaningful and sustainable ways. Grounded studies contribute pivotal insights into how ethnography *with* people in 'lived' energy (crisis) situations, can help us research future imaginations. Perhaps most important of all, as underscored by Star (1999: 379) is what values and ethical principles we inscribe in the inner depths of the built information environment. We need methods to understand this imbrication of infrastructure and human organisation.

Studying energy futures is an emerging field that opens analytical and practical possibilities for imagining and practising futures otherwise, forging new forms of collaboration, and producing 'spaces of possibilities' through which descriptions of partial (im)possibilities can be generated and embraced. The ethnographic register can make a contribution in analysing how energy futures are performed in specific contexts (Watts 2018), which logics are reproduced, and which ones can be challenged. In contrast to Ch. 2, the empirical cases we draw from in this chapter engage with groups of people who are to a large degree dispossessed by the ways in which energy futures in their environment are conjured up and acted on/towards. That chapter engaged ethnographically with people in affluent nations (Australia, the United Kingdom, and Europe) whose access to everyday energy is predominantly via electricity grids, fuel stations, and consumer products, and shaped by the techno-solutionist agenda of neoliberal governments and powerful industry stakeholders. Its authors seek to complicate dominant narratives about energy and technology futures proposed by consultancies, industry, and policy bodies, from the very sites that these future visions superficially appear to be consistent with. Our chapter also seeks to challenge such narratives, but with differently situated groups of people. Social norms and structures shape what technological tools are imagined as necessary in the first place, just as the connecting infrastructure reaches the North first, and then the affluent segments of the population in the South (Anand, Gupta, and Appel 2018). We should never disregard the specific colonial continuities of technological change and energy infrastructure, but instead practice a simultaneous sensibility to the alternative ways of perceiving energy and the environment locally, and the continued resistance and continued hope of other energy futures – futures that do not treat human, non-human, and earth as standing reserves. This chapter has presented initial steps, through grounded engaged ethnography, of imagining energy futures beyond colonial continuities and be part of articulating this, even at times as part of enunciatory communities.

A futures anthropology strives to be decolonial in that it considers how the research will circulate and is part of forging futures, striving to be accountable in terms of how the knowledge we make comes to matter and for whom: Who will access it? In what form? Whom does it benefit and how? This entails working multimodally both methodologically and in disseminating our knowledge to different groups and publics, as we continually interrogate the god-eye trick (Haraway 1988), thinking that an objective, detached knowledge-making that does not account for itself is possible or desirable. It also entails making space for approaches, and modalities that attend to the relational–Indigenous arts practice as resistance to extractivism, for instance, or knowledge practices such as tracking as modes of noticing colonial or extractive violences – but are so often marginalised by dominant logics.

All authors wish to thank the people who spent time with them in their respective fieldsites – without whom the research and knowledge-generation would not be possible.

Karen Waltorp wishes to acknowledge a grant from the Independent Research Fund Denmark no. 1130 – 00019B 'Digital everyday lives far from Silicon Valley: Technological Imaginaries and Energy Futures in a South African Township (DigiSAt)'.

Martín Fonck wishes to acknowledge the grant from the program 'Beca de Doctorado en el Extranjero Becas Chile' from the Chilean National Agency for Research and Development. He also wishes to thank the program 'Global Cultures – Connecting Worlds', the Institute of Social and Cultural Anthropology, the Rachel Carson Center in Environment and Society at the University of Munich, and the Institute for Advanced Sustainability Studies. Finally, he wants to thank the support during fieldwork of the Andean Geothermal Center of Excellence.

Pierre du Plessis wishes to acknowledge the International Postdoctoral grant from the Independent Research fund Denmark no. 9032 – 00012B 'Enacting Contested Landscapes: Dwelling, Conservation, and Prospecting in the Kalahari Desert'. He is also deeply grateful for support from Environmental Humanities South at the University of Cape Town and the Anthropology Department at Aarhus University for co-hosting this research period.

Ragnhild Freng Dale wishes to acknowledge a PhD grant from the Gates Cambridge Scholarship programme at the University of Cambridge and support for writing by the Norwegian Research Council grant no. 296205 (FME NTRANS). Most importantly, she is grateful for the unwavering strength of Máret Ánne and Jovsset Ánte Sara's work which inspired her part of this chapter.

References

Abram, S., Winthereik, B. R., and Yarrow, T. (eds.) (2019). *Electrifying anthropology: exploring electrical practices and infrastructures*. London: Bloomsbury Academic.

Adhikari, M. (2006). 'God made the white man, God made the black man…': popular racial stereotyping of Coloured people in apartheid South Africa. *South African historical journal*, 55, 142–164.

Allen, J. S., and Jobson, R. C. (2016). The decolonizing generation: (race and) theory in anthropology since the eighties. *Current anthropology*, 57(2), 129–148.

Anand, N., Gupta A., and Appel, H. (2018). *The promise of infrastructure*. Durham, RC: Duke University Press.

Barbee, J. (2015, December 2). Botswana sells fracking rights in national park. *The Guardian*. https://www.theguardian.com/environment/2015/dec/02/botswana-sells-fracking-rights-in-national-park

Barbee, J., Dutschke, M., and Smith, D. (2013). Botswana faces questions over licences for fracking companies in Kalahari. *The Guardian*. https://www.theguardian.com/environment/2013/nov/18/botswana-accusations-fracking-kalahari

Bauer, C. (1998). *Against the current: privatization, water markets, and the state in Chile*. Boston: Kluwer Academic Publishers.

Benjamin, R. (2019). Race after technology: abolitionist tools for the new Jim code. Hoboken NJ: John Wiley and Sons.

Biko, S. (2002/1978). *I write what I like: selected writings*. Chicago: University of Chicago Press.

Bjørklund, I. (2016). Fra formynder til forhandler: om inngrep, konsekvensanalyser og 'balansert sameksistens'. In T. A. Benjaminsen, I. M. G. Eira, and M. N. Sara (eds.), *Samisk reindrift, norske myter*. Bergen: Fagbokforlaget. 177–194.

Bonelli, C., and Dorador, C. (2021). Endangered salares: micro-disasters in Northern Chile. *Tapuya: Latin American science, technology and society*, 4(1), 1–29. https://doi.org/10.1080/25729861.2021.1968634.

Bowker, G. C., and Star, S. L. (1999). *Sorting things out: classification and its consequences*. Cambridge, MA: MIT Press.

Brüggen, J. (1940). *Informe geológico sobre los géiseres Volcanes del Tatio*, Ministerio de Fomento, Departamento de Minas y Petróleo.

Carrasco, A. (2020). *Embracing the anaconda: a chronicle of Atacameño life and mining in the Andes*. Lanham: Lexington Books.

Césaire, A. (2000). *Discourse on colonialism*. New York: Monthly Review Press.

Dale, R. F. (2018). *Making resource futures: petroleum and performance by the Norwegian Barents Sea*. University of Cambridge.

Dankertsen, A. (2016). Fragments of the future. Decolonization in Sami everyday life. *Kult*, 14, 23–37.

Decreto Exento N° 667. (2014, August 13) *Otorga concesión gratuita de inmueble fiscal en la región de Antofagasta, a comunidad atacameña de Toconce y a comunidad atacameña de Caspana.* Ministerio de Bienes Nacionales. Diario Oficial de la República de Chile, Santiago, Chile.

Du Plessis, P. L. (2010). *Tracking knowledge: science, tracking and technology* [Thesis, University of Cape Town]. https://open.uct.ac.za/handle/11427/14263.

Du Plessis, P. L. (2018). *Gathering the Kalahari: tracking landscapes in motion.* PhD Dissertation. Aarhus University/University of California, Santa Cruz.

Du Plessis, P. L. (2022). Tracking meat of the sand: noticing multispecies landscapes in the Kalahari. *Environmental Humanities*, 14(1), 49–70.

Erasmus, Z. (2017). *Race otherwise: forging a new humanism for South Africa.* Johannesburg: Wits University Press.

Escobar, A. (2007). Worlds and knowledges otherwise. *Cultural studies*, 21(2–3), 179–210. DOI: 10.1080/09502380601162506.

Fanon, F. (1963). *The wretched of the earth.* New York: Grove Press.

Fanon, F. (1967). *A dying colonialism.* New York: Grove Press.

Fanon, F. (2008). *Black skin, white masks.* (New edn.) New York: Grove Press.

Fonck, M. (2021). Subterranean (in)visibilities: traces, underground water, and thermal flows in the El Tatio Geyser Field, Atacama, Chile. Environment and Society Portal, *Arcadia*. Rachel Carson Center for Environment and Society. doi:10.5282/rcc/9316.

Fortun, K. (2001). *Advocacy after Bhopal: environmentalism, disaster, new global orders.* Chicago: University of Chicago Press.

Glissant, E. (1997). *Poetics of relation.* Ann Arbor: University of Michigan Press.

Hall, S., Morley, D., and Chen, K.-H. (1996). *Stuart Hall: critical dialogues in cultural studies.* London: Routledge.

Haraway, D. (1988). Situated knowledges: the science question in feminism and the privilege of partial perspective. *Feminist studies*, 14(3), 575–599.

Harrison, F. V. (2011). *Decolonizing anthropology: moving further toward an anthropology for liberation* (3rd edn.). American Anthropological Association.

Heidegger, M. (1977). *The question concerning technology and other essays.* New York: Harper and Row.

Henriksen, G. (2003). Consultancy and advocacy as radical anthropology. *Social analysis: the international journal of social and cultural practice*, 47(1), 116–123.

High, M. M., and Smith, J. M. (2019). Introduction: the ethical constitution of energy dilemmas. *Journal of the Royal Anthropological Institute*, 25, 9–28.

Holmberg. A. (2018) Bivdit Luosa – To Ask for Salmon. Saami Traditional Knowledge on Salmon and the River Deatnu: In Research and Decision-Making. MA Thesis, UiT The Arctic University of Norway.

Hu, T.-H. (2015). *A prehistory of the cloud.* Cambridge MA: MIT Press.

Jobson, R. C. (2020). The case for letting anthropology burn: sociocultural anthropology in 2019. *American anthropologist*, 122(2), 259–271.

Johnsen, K. I., Benjaminsen, T. A., and Eira, I. M. G. (2015). Seeing like the state or like pastoralists? Conflicting narratives on the governance of Sámi reindeer husbandry in

Finnmark, Norway. *Norsk geografisk tidsskrift – Norwegian journal of geography*, 69(4), 230–241.

Keeping, D., Kashe, N., Langwane, H. (Karoha), Sebati, P., Brahman, !N., Xhukwe, Q., Molese, N., Gielen, M.-C., and Keitsilebarunggwi, A. (2019). Botswana's wildlife losing ground as Kalahari Wildlife Management Areas (WMAs) are dezoned for livestock expansion. *BioRxiv*. https://doi.org/10.1101/576496.

Kelley, R. D. G. (2000). A poetics of anticolonialism. In A. Césaire, *Discourse on colonialism*. New York: Monthly Review Press. 7–28.

Kuokkanen, R. (2000). Towards an Indigenous paradigm from a Sami perspective. *The Canadian journal of native studies*, 20(2), 411–436.

Larkin, B. (2013). The politics and poetics of infrastructure. *Annual review of anthropology*, 42, 327–343.

Latour, B. (2004). Why has critique run out of steam? From matters of fact to matters of concern. *Critical inquiry*, 30 (2), 225–248.

Ley N°19.657. (2000). Sobre concesiones de energía geotérmica. *Diario Oficial de la República de Chile*, Santiago, Chile.

Livingston, J. (2019). *Self-devouring growth: a planetary parable as told from Southern Africa*. Durham, NC: Duke University Press.

Magubane, M. (1979). *The political economy of race and class in South Africa*. New York and London: Monthly Review Press.

Marres, N. (2005). Issues spark a public into being: a key but often forgotten point of the Lippmann-Dewey debate. In B. Latour and P. Weibel (eds.), *Making things public*. Cambridge, MA: MIT Press. 208–217.

McTighe, L., and Raschig, M. (2019). An otherwise anthropology. Theorizing the contemporary, fieldsights. *Cultural anthropology*. https://culanth.org/fieldsights/series/an-otherwise-anthropology.

Mendez, M., Prieto, M., and Godoy, M. (2020). Production of subterranean resources in the Atacama Desert: 19th and early 20th century mining/water extraction in the Taltal district, northern Chile. *Political Geography*, 81, 102194. https://doi.org/10.1016/j.pol geo.2020.102194.

Minh-ha, T. T. (1990). *Woman, native, other: writing postcoloniality and feminism*. Bloomington: Indiana University Press.

Mohanty, C. (1988). Under Western eyes: feminist scholarship and colonial discourses. *Feminist review*, 30, 61–88. https://doi.org/10.1057/fr.1988.42.

Morata, D. (2014). '¿Chile: un país geotérmico en un futuro inmediato?', *Anales de la Universidad de Chile*, 5 June 71–86. doi:10.5354/0717-8883.2014.31635.

Nixon, R. (2011). *Slow violence and the environmentalism of the poor*. Cambridge, MA: Harvard University Press.

Normann, S. (2021). Green colonialism in the Nordic context: exploring Southern Saami representations of wind energy development. *Journal of community psychology*, 49(1), 77–94.

Otte, P., Rønningen, K., and Moe, E. (2018). Contested wind energy: discourses on energy impacts and their significance for energy justice in Fosen. In A. Szolucha (ed.), *Energy, resource extraction and society*. Abingdon: Routledge. 140–158.

Prieto, M., and Bauer, C. (2012). Hydroelectric power generation in Chile: an institutional critique of the neutrality of market mechanisms. *Water international*, 37, 131–146. https://doi.org/10.1080/02508060.2012.662731.

Reinert, H. (2019). The Skulls and the Dancing Pig. *Terrain*, 71. https://doi.org/10.4000/terrain.18051.

Reuters (2021). Botswana signs 10 MW gas-fired deal with Tlou to wean off coal, imports. https://www.reuters.com/business/energy/botswana-signs-10-mw-gas-fired-deal-with-tlou-wean-off-coal-imports-2021-10-18/

Robins, S, (2019). 'Day Zero', hydraulic citizenship and the defence of the commons in Cape Town: a case study of the politics of water and its infratructure (2017–2018). *Journal of southern African studies*, 45(1), 5–29.

Sandström, M. (2020). *Dekoloniseringskonst: artivism i 2010-talets Sápmi – Decolonising artivism in contemporary Sápmi*. Thesis, Department of language studies, Faculty of Arts, Umeå University: Umeå universitet.

Simone, A. (2004). People as infrastructure: intersecting fragments in Johannesburg. *Public culture* 16(3), 407–429.

Spivak, G. (1988). Can the subaltern speak? In C. Nelson and L. Grossberg (eds.), *Marxism and the interpretation of culture*. Urbana: University of Illinois Press. 271–313.

Staff Reporter (2013). Botswana government rejects fracking claims. *The mail and guardian*. https://mg.co.za/article/2013-11-20-botswana-government-rejects-fracking-claims/

Star, S. L. (1999). The ethnography of infrastructure. *American behavioral scientist*, 43(3), 377–391. https://doi.org/10.1177/00027649921955326.

Stats SA (2012). *Census 2011 Municipal report: Western Cape*. Statistics South Africa.

Stephansen, M. (2017). A hand-drawn map as a decolonising document: Keviselie (Hans Ragnar Mathisen) and the artistic empowerment of the Sami movement. *Afterall: a journal of art, context and enquiry*, 44, 112–121.

Stock Market Wire (2015). *Nodding Donkey awarded Botswana licences*. Stock Market Wire. https://www.stockmarketwire.com/article/5105033/Nodding-Donkey-awarded-Botswana-licences.html.

Strathern, M., Sasser, J. S., Clarke, A., Benjamin, R., TallBear, K., Murphy, M., Haraway, D., Huang, Y.-L., and Wu, C.-L. (2019). Forum on making kin not population: reconceiving generations. *Feminist studies*, 45(1), 159–172.

Strengers, Y. (2013). *Smart energy technologies in everyday life. Smart utopia?* London: Palgrave Macmillan.

TallBear, K. (2013). Genomic articulations of indigeneity. *Social studies of science*, 43(4), 509–534.

Tocchi, E. (1923). *Il Tatio, Ufficio geologico Larderello SpA*. Unpublished report.

Tsing, A. L. (2015). *The mushroom at the end of the world: on the possibility of life in capitalist ruins*. Princeton, NJ: Princeton University Press.

Vistnes, I. I., Burgess, P., Mathiesen, S., Nellemann, C., Oskal, A., and Turi, J. M. (2009). *Reindeer husbandry and Barents 2030: impacts of future petroleum development on reindeer husbandry in the Barents region*. Alta: International Centre for Reindeer Husbandry.

Waltorp, K. (2018). Fieldwork as interface: digital technologies, moral worlds and zones of encounter. In A. Estalella and T. Sánchez Criado (eds.), *Experimental collaborations: ethnography through fieldwork devices*. London: Berghahn Books. 114–131.

Waltorp, K. (2020). *Why Muslim women and smartphones: mirror images*. London: Routledge.

Waltorp, K. (2021). Multimodal sorting: the flow of images across social media and anthropological analysis. In A. Ballestero and B. R. Winthereik (eds.), *Experimenting with ethnography*. Durham, NC: Duke University Press. 133–150.

Waltorp, K., and ARTlife Film Collective (2021). Isomorphic articulations: notes from collaborative film-work in an Afghan–Danish film collective. In L. Di Puppo, L. Martínez, and M. D. Frederiksen, *Peripheral methodologies: unlearning, not-knowing and ethnographic limits* (Anthropological Studies of Creativity and Perception). London: Routledge. 115–130.

Waltorp, K., and Jensen, S. (2019). Awkward entanglements: kinship, morality and survival in Cape Town's prison-township circuit. *Ethnos*, 84(1), 41–55.

Waltorp, K., and Vium, C. (2010). *Manenberg: growing up in the shadows of apartheid*, (film), distributed by DR International sales/Royal Anthropological Institute.

Waltorp, K., Lanzeni, D., Pink, S., and Smith, R. C. (forthcoming). Introduction: an anthropology of futures and technologies. *An anthropology of futures and technologies*. London and New York: Routledge.

Watts, L. (2018). *Energy at the end of the world: an Orkney Islands saga*. Cambridge, MA: MIT Press.

Wilkie, A., Savransky, M., and Rosengarten, M. (2017). *Speculative research: the lure of possible futures*. London; New York: Taylor and Francis.

Author biographies

Simone Abram is Professor in Anthropology at Durham University UK and is a co-director of the Durham Energy Institute. She co-founded the EASA Energy Anthropology Network with Nathalie Ortar in 2016. She is currently Chair of the Association of Social Anthropologists of the UK (ASA). Publications include *Ethnographies of power* (2020), *Electrifying anthropology* (2019), *Green ice* (2016), *Media, engagement and anthropological practice* (2015), *Elusive promises* (2013) and *Culture and planning* (2011).

Chiara Bresciani is an anthropologist based at Aarhus University, Denmark where she teaches at the MA in International Studies. Her PhD focuses on social change and the impact of a wind energy project on tradition, heritage and identity in an indigenous village on the coast of Southern Mexico, where she has been conducting ethnographic fieldwork since 2011. As an ethnographer, she is also a consultant in a study on postwar economic reconstruction for a collaborative project on memory of conflict in El Salvador.

Patrick Brodie is a media scholar and FRQSC Postdoctoral Fellow in the Department of Art History and Communication Studies at McGill University. His research unravels the cultural and environmental politics of digital media infrastructures, with a specific focus on extractivism and rural communities. His work has appeared and is forthcoming in *Media, culture and society*, *New media and society*, *Information, communication and society*, *Environment and planning E: nature and space*, and *Canadian journal of communication*, among other venues.

Liza Rose Cirolia is a senior researcher at the African Centre for Cities, University of Cape Town (South Africa). Her work focuses on governance, municipal finance, urban infrastructure, and real estate in the context of Africa's growing urban areas. Her recent projects focus on distributed and decentralized infrastructure configurations, taking an approach that aims to be both critical and propositional. She is actively involved in developing pan-African research collaborations, leading collaborative and comparative projects. In addition to academic work, Liza consults for local and international organizations, supporting policy development processes in African cities.

Kari Dahlgren is a Research Fellow in the Emerging Technologies Research Lab at Monash University. She is an anthropologist and ethnographer who studies the social dynamics of energy consumption and production in Australia. In the context of increasing questions about the future of energy and the environment, Kari studies the everyday moral and ethical implications of transition for the energy and extractive industries. She holds a PhD from the London School of Economics and an MSc from Oxford University.

Ragnhild Freng Dale is a social anthropologist and senior researcher at the Western Norway Research Institute. Her research interests include energy, infrastructure and their related narratives, community impacts, indigenous perspectives and environmental activism, primarily in Sápmi and Norway. She has also carried out a number of projects in collaboration with artists and the performing arts field. She holds a PhD from the Scott Polar Research Institute at the University of Cambridge, and is a member of the Young Academy of Norway.

https://doi.org/10.1515/9783110745641-007

Pierre du Plessis is an environmental anthropologist from Botswana whose research has broadly focused on human–environmental relations and multispecies ethnography in the Kalahari Desert. His research has explored the practices of tracking and gathering as modes of noticing, describing, and theorizing landscapes and landscape change. He is currently a Researcher at the Oslo School of Environmental Humanities, University of Oslo, and previously held the Independent Research Fund Denmark-funded International Postdoctoral Grant, for which he was co-hosted by the Anthropology Department/Centre for Environmental Humanities at Aarhus University and the Environmental Humanities South program at the University of Cape Town (2019–2021).

Katherine Ellsworth-Krebs is a Senior Research Associate in Sustainability at the Imagination Lancaster Laboratory, Lancaster University and she was previously a Lecturer in Sustainable Development at the University of St Andrews. Her research focuses on how 'normal' expectations of home are becoming increasingly energy-demanding and much of her work aims to bring the wealth of scholarship on the meaning and making of home into energy debates.

Aurore Flipo is a sociologist and associate researcher at PACTE-University of Grenoble-Alpes. Her research has been focusing on mobility as a socially stratified practice of negotiating spatial resources and constraints. After a PhD on labour migration in the European Union (published by Rennes University Press), she has been researching the interactions between daily mobility, labour and residential mobility within a research project on remote working in coworking spaces, with Nathalie Ortar and Patricia Lejoux. Since 2018, she has focused on the specificities of rural areas and the issue of social change and sustainability in energy practices.

Martín Fonck is a research associate at the Institute of Advanced Studies in Sustainability in Potsdam. He completed his PhD at the Rachel Carson Center in Environment and Society and the Institute of Social and Cultural Anthropology, University of Munich. His main interest is studying narratives and imaginaries of environmental politics in the transition towards sustainability. Currently, he is writing about technologies of exploration based on his doctoral project on geothermal energy futures in the Chilean Andes. He is part of the IASS research group Planetary Geopolitics and Geoengineering, studying how nature is transformed in unexpected technologies and the imaginaries of future landscapes.

Alix Johnson is a Cultural Anthropologist and Assistant Professor of International Studies at Macalester College. Her work, which has appeared in *American ethnologist*, *American anthropologist*, *City and society* and *Culture machine*, takes up technological infrastructures as a lens on questions of sovereignty, emerging spatial politics, and enduring formations of imperial power – especially in Iceland and the broader Arctic.

Michiel Köhne is assistant professor at Wageningen University. His research topics include conflicts around oil palm plantations in Indonesia, resistance against coal mining and coal seam gas and shale gas extraction in Australia and the Netherlands and the energy transition. His present field work focuses on activism among scientists.

Hsin-yi Lu is an associate professor of anthropology at National Taiwan University. Her research interests include political ecology, anthropology of space and place, energy politics, and artisan fishery practices in Taiwan. Recent works are published in *Taiwan journal of anthropology* (2020), *Journal of anthropology and archeology* (2021), and *Journal of Chinese dietary culture* (2019). Her current research project 'Infrastructuring the ocean: an ethnography of Taiwan's offshore wind development' aims to explicate the spatial, temporal, and material politics involved in the planning and installation processes of Taiwan's offshore wind farms.

Clément Marquet is a postdoctoral student in sociology at the Costech laboratory of the University of Technology of Compiègne, on a grant from the Institut Francilien Recherche Innovation Société (IFRIS). His PhD focused on the development of the city through informational (applications, data) and physical (data centres) digital infrastructures. In his current research, he investigates the transformations of digital technology in the context of climate emergency, by studying both the controversies surrounding the evaluation of the environmental footprint of the technological system and the ecological, political, and social stakes of infrastructure deployment.

Katja Müller conducts research into digitisation, museum studies, material culture and visual anthropology, as well as energy and environmental humanities. She is Visiting Professor at the University of Technology Sydney, and Privatdozentin for social anthropology at Martin Luther University Halle-Wittenberg. Her latest books include *Beyond the coal rush* and *Digital archives and collections*, analysing the coal rush in Germany, Australia and India and online access to heritage material in India and Europe.

Nathalie Ortar is Senior Researcher in anthropology at the ENTPE-University of Lyon. Her research has mainly focused on energy, refuse, routines, housing, and spatial mobility. She co-founded the EASA Energy Anthropology Network with Simone Abram in 2016. Publications include *Becoming urban cyclists* (2022), *Ethnographies of Power* (2021), *L'énergie et ses usages domestiques* (2018), *Migrations, circulations, mobilités* (2018), *Jeux de pouvoir dans nos poubelles* (2017), and *La deuxième vie des objets: Recyclage et récupération dans les sociétés contemporaines* (2015).

Sarah Pink is Professor and Director of the Emerging Technologies Research Lab and Associate Director of Monash Energy Institute at Monash University, and CI in the ARC Center of Excellence for Automated Decision-Making and Society. In 2014, she co-founded the Future Anthropologies Network of the EASA with Juan F. Salazar. Sarah's recent publications, all in 2022, include three books – *Emerging technologies/life at the edge of the future*, *Design ethnography*, and *Everyday automation* – and the documentary film *Digital energy futures*.

Andrea Pollio is Marie Skłodowska-Curie Research Fellow jointly at the African Centre for Cities, University of Cape Town, and at the department of Urban and Regional Studies and Planning (DIST) at the Polytechnic of Turin. His research interests are situated at the intersection of STS, urban studies and development economics, and his published works explore the interface between technology economies, development and urbanization in Africa. Andrea's current research project is funded by a Marie Skłodowska-Curie fellowship (grant 886772), and addresses the impact of private Chinese technology companies on Nairobi's Silicon Savannah.

Elisabet Dueholm Rasch is an associate professor at Wageningen University. Her research topics include (indigenous) mobilization against neoliberal policies and extractive projects, and energy production in Latin America (Guatemala) and the Netherlands. Her contemporary fieldwork in Guatemala focuses on how territory defenders experience violence and criminalization.

Antti Silvast is an Associate Professor at the Technical University of Denmark, DTU Management, Division for Responsible Innovation and Design. He is a sociologist examining energy infrastructure, smart systems, methodology, and energy modelling. He is an editor of the journal *Science and technology studies* and has published 35 peer-reviewed articles and chapters, and two monographs. He held postdoctoral positions at Princeton University (Princeton Institute for International and Regional Studies), Durham University (Department of Anthropology), University of Edinburgh (Science, Technology and Innovation Studies), and the Norwegian University of Science and Technology. He received his PhD from the University of Helsinki (Sociology, 2013).

Alexander R. E. Taylor is a social anthropologist and Lecturer in Communications at the University of Exeter. He is also the Marconi Fellow in the History and Science of Wireless Communication at the University of Oxford. He works at the intersection of digital anthropology, infrastructure studies and media and communication studies. His research concentrates on the materiality and security of communications infrastructure. He is an Editorial Assistant for the *Journal of extreme anthropology* and a founder of the Cambridge Infrastructure Resilience Group, a network of researchers exploring critical infrastructure protection in relation to global catastrophic risks.

Julia Velkova is associate professor of media and communication studies at Linköping University, Sweden. Her work on the temporalities, energy politics, and human work in data centres has been published in journals such as *New media and society*, *Big data and society*, and *Information, communication and society* among other venues. She is currently involved in four research projects that concern different intersections of energy and digitalisation from a sociotechnical perspective. She is also co-editing the book *Media backends: critical studies of the other side of the screen* together with Lisa Parks and Sander De Ridder (forthcoming with University of Illinois Press).

Asta Vonderau is Professor of Anthropology and co-director of the Centre for Interdisciplinary Regional Studies and Centre for Just Transition and Sustainability at the University of Halle-Wittenberg, Germany. She co-heads the City Industries Research Network together with Kerstin Pöhls. Her research focuses on anthropology of policy, energy, and resources as well as research methodologies in/of data saturated worlds. Vonderau's work has appeared in *Ethnos*, *Culture machine*, *Journal of cross-cultural image studies*, and *International journal of higher education in the social sciences*. Publications include *The nature of data centers* (2019), *Formationen des Politischen* (2014), and *Changing economies, changing identities in postsocialist Eastern Europe* (2008).

Karen Waltorp is associate professor and filmmaker and Head of the Ethnographic Exploratory, Department of Anthropology, University of Copenhagen. She coordinates the Researcher Group Technē, the EASA Future Anthropologies Network, and serves on the Editorial Board of

Cultural Anthropology. Waltorp is the author of *Why Muslim Women and Smartphones* (Routledge 2020) and has published widely on gender, migration, and new technologies, multimodal and digital methodologies. She was Co-PI on ARTlife, a 'research-through-collaborative-filmmaking' project selected for Do:Lab 2022–23, and is currently PI on DigiSAt: Digital Everyday Lives far from Silicon Valley (DigiSAtproject.com).

Index

Note: Page references in *Italic* refer to *figures*

https://doi.org/10.1515/9783110745641-008

www.ingramcontent.com/pod-product-compliance
Lightning Source LLC
Chambersburg PA
CBHW020530270326
41927CB00006B/523